Work, Community, and the Mining Wars
in the Central Illinois Coal Fields During the Great Depression

Divided Kingdom

Second Edition

Carl D. Oblinger

Illinois State Historical Society
Springfield, Illinois
2004

Divided Kingdom

This volume was published by the Illinois State Historical Society with the cooperation and support of the Illinois Historic Preservation Agency in its continuing effort to preserve the history of the people of the state. The first edition was funded in part by a grant from the Illinois Humanities Council, a statewide organization funded in part by the National Endowment for the Humanities. The ISHS provided a matching grant for that publication. The Southern Illinois University Coal Research Center provided a grant for publishing the 2nd edition, as did the Illinois State Historical Society. The author provided funding for the initial design and copy. Opinions and ideas expressed in the second edition do not necessarily represent the views of the Illinois Historic Preservation Agency, the Illinois State Historical Society, the Illinois Humanities Council, the Coal Research Center or any agency of the federal or state government.

Copyright © 2004, © 1991 by the Illinois State Historical Society, Springfield, Illinois. All rights reserved.
Cover & book design by Charles J. Copley Graphics

Library of Congress Catalog Card Number: 91-076646

Cover Photo: Anonymous miner from Hewetville section of Taylorville posing with his pride. According to retired miner Fred Boch, who furnished the majority of photos used in this book, the miner made his money in mining and moved back to Italy before World War I.

All photographs were reproduced courtesy of the Abraham Lincoln Presidential Library, a division of the Illinois Historic Preservation Agency. Mary Michals, Iconographer.

To the Boch Family
For their contribution to the preservation
of the history of the Midland Tract

and to the memory of
Senator Vince Demuzio
The Late Majority Leader and
Senator of the 49th Senatorial
District of Illinois

Divided Kingdom

Just Scabs

by Mrs. Mary Libby

Old Taylorville has lots of scabs
And Tovey just a few,
But it keeps them busy picking scabs,
They have nothing else to do.

There was once a grand old union,
But it turned to a dirty scab,
And so the wise men left it,
And the scabs were left to scab.

The man who wears the name of scab,
Who works behind a gun,
Is the foulest, slimy, creeping thing,
That crawls beneath the sun.

Old scabs, young scabs,
You can smell them in the air,
At Taylorville and Tovey,
You can find them everywhere!

Oh, Taylorville and Tovey,
It's a time to close your door,
'Til every louse kills every scab,
And be clean, free men once more.

The Progressive Miner, November 11, 1932

Contents

List of Figures vi
Preface 1
Assumptions & Acknowledgments 3
Methodology 8
List of Miners Interviewed 13
Introduction 14
A Chronology of the Establishment of
Midland Tract Towns and Mines 36
Chapter I. Family & Community: Sinking New Roots ... 39
    British Immigrants 46
    Southern and Eastern Europe Immigrants 53
    Family 65
    Community 81
Chapter II. Work, Safety, and the United Mine Workers:
The World of Underground Mines 95
    Early Mining and Mechanization 95
    Company Practices and Safety 134
    Union Traditions & the United Mine Workers 154
Chapter III. The Miners' War: Defending the Miners' World 163
A Chronology of Events 166
    Causes 173
    The Progressive Miners 187
    Strikebreakers 205
    The Violence 235
    Afterward 246
Conclusion 249
Appendices 254
    Glossary 255
    Questionnaire 260
Bibliography 267

Divided Kingdom

Figures

1. Ferdinand Boch, 1911. ix
2. The Interviewing Team..7
3. *Daily Breeze* office, Taylorville, Sept. 29, 1932..................17
4. Illinois National Guard Officers, Fall, 1932........................17
5. Location of Peabody Coal Mines and Towns: Christian County, 1932..18
6. Peabody Coal, Mine No. 9, 1917......................................20
7. C&IM Water Tank, 1913..20
8. Miners on Strike, Peabody Mine No. 58, Aug. 1932.........29
9. Picket at Mine No. 9, August, 1932................................29
10. Community Solidarity, October, 1932............................34
11. Food Distribution Station, Taylorville, 1932...................34
12. Boch Family..40
13. Young Rock Pickers on Picnic, 1926...............................44
14. Christian County Coal Mine, early 1900s.......................47
15. Mines of Christian County Coal Mine, early 1900s........47
16. Stephen Boch House, Hewittville, 1929.........................57
17. Peabody Mine No. 58, early 1930s.................................57
18. Peabody Mine No. 9, Langley, 1917...............................62
19. Typical Coal Miner's Housing, Moweaqua, 1932...........66
20. Frank Boch in strawberry patch, 1911............................75
21. Boch Brothers, with strawberries, 1914..........................75
22. The Kondike Hotel, Taylorville, 1925.............................77
23. Troops in Manner's Park, Taylorville, 1932...................84

24. Baseball Game in Pawnee, 1925 85
25. Star Trading Company, 1904 85
26. Rock Pickers' Picnic, 1926 87
27. The Glove Factory, Taylorville, 1920s 89
28. Stonington Concert Band, 1920s 89
29. Merchant's Band, Pawnee, 1928 91
30. P.M.A. Local No. 1, Miners' Band, 1937 91
31. Timbermen, Mine No. 58, 1920s 96
32. Drillers in Mine No. 58, 1920s 99
33. Miners and Helpers, Mine No. 58, 1920s 99
34. Buggy and Miners, Mine No. 7, 1916 100
35. Buggies with Flat Coal, Mine No. 58, 1920s 100
36. Coal Buggies at Moweaqua Coal Mine, 1933 101
37. Trip Rider at Mine No. 9, 1930 101
38. Cutting Machine Underground, 1920s 102
39. Cutting Machine at the Face, 1920s 102
40. Loading Machine, 1930s 104
41. Mule and Coal Buggies, Mine No. 9, 1917 105
42. Employees of the Christian County Coal Co., 1905 105
43. Road Motor, Mine No. 58, 1920s 107
44. Entry Gang, Mine No. 58, 1920s 109
45. Blacksmith Shop, Christian County Coal Company, 1906 109
46. Dirt Gang, Mine No. 58, 1940s 112
47. Coal Loaders, Mine No. 58, 1930s 114
48. Gang on Cutting Machine, Mine No. 58, 1930s 120
49. William Augustus Pierce, old miner, 1923 134
50. Accident on C&IM Railroad, Pawnee, 1908 145
51. Underground Miners, Mine No. 58, late 1930s 151
52. Trip in Moweaqua Coal Company Mine, 1933 153
53. Progressive Mines of America, Taylorville office, March, 1933 163
54. Labor Day Parade, Springfield, 1935 164
55. Peabody Mine No. 5, Dismantled, 1921 176
56. Caravan of Miners, August, 1932 182
57. Road Blocked by Miners, August, 1932 184
58. National Guard Troops in Taylorville, October 12, 1932 184
59. Miners in Christian County Court House, October, 1932 187

Divided Kingdom

60. Courtroom, Christian County Court House, October, 1932..........188
61. Strikers at Peabody's Mine No. 7, August 1932..............188
62. National Guard Troops quartered in Jury Room, 1929.................192
63. "Heating coffee," Taylorville, August, 1932......................201
64. Striking Miners outside Mine No. 58, August, 1932......................224
65. Striking Miners at Mass Meeting, Mine No. 9, August, 1932.........235
66. Funeral in Taylorville, October 16, 1932..........................235
67. Bombed-out *Daily Breeze* offices, Sept. 28, 1932 239
68. Interior of the *Breeze* Building, September 28, 1932....................239
69. Miners' Rally, Kincaid, August 1932................................243
70. Miners Sleeping at Rally, Taylorville, August, 1932......................243
71. Andrew Gyenes Funeral, Manner's Park, Taylorville, October 16, 1932..243
72. Casket of Andrew Gyenes, Manner's Park, Taylorville, October 16, 1932..246
73. Opening of Peabody Coal Mine No. 10, June 13, 1951 252

Figure 1. Ferdinand Boch in a picture taken about 1911. The Boch brothers—Ferdinand, Fred, and Max—allowed the interviewers easy access to the large collection of photos accumulated by their family over a hundred-year career of amateur photography. (Courtesy of the Boch Brothers)

Divided Kingdom

Preface

The coal fields of central Illinois formed the backbone of a strong regional economy for the first half of the 20th Century. These coal mines shaped the identity of society within the region. The values of those families and communities are still exhibited today in their work ethic and family ideals.

This revision of *Divided Kingdom* again defines the struggles that shaped and formed the lives and society of central Illinois. The story of these people is more than just an economic struggle or even a struggle between competing unions; it shows a way of life that developed based upon democratic principles and comradery unparalled in modern day American history.

The struggles encountered in the central Illinois coal fields in the 1930s, and still relived today, have been largely ignored and certainly misunderstood. The general decline of the central Illinois rural economy owes its beginning to the lost strike of the rank and file miners in the 1930s. Communities entirely dependent on coal mining lost a vital core of activist miners and entrepreneurs by the late 1930s as many of the mines closed.

The words of those who firsthand experienced the struggles and fully participated in the events of the time, are rich

Divided Kingdom

in meaning and illustrative of the character-shaping forces that molded their society and left a history of which we can be proud. I should know; my father was a Progressive miner who struggled for the rights of all working men and women in the Macoupin County mines during this period.

We owe a great debt to all of those who struggled and endured hardships. They have taught us by their example some of the greatest lessons of life.

To Carl Oblinger and all the others who helped assemble this work, we owe another debt to bringing to our attention the forces which have shaped our past and help explain the present.

Senator Vince Demuzio
The Late Majority Leader and
Senator of the 49th Senatorial
District of Illinois

Divided Kingdom

Assumptions & Acknowledgments

The original publication of *Divided Kingdom* occurred thirteen years ago. It was based on an interpretation of miners' and their wives' recollections of a strike in Central Illinois during the Great Depression. It was also an attempt to let these workers speak for themselves. They had been forgotten for so long not because they were silent, but because their explanations of their struggles were not valued as they should have been.

The book is based on the belief that the social history of a people in a given historical epoch must begin with the testimony of the people themselves. "If you want the miners' history," a striking coal miner told Mother Jones in Cabin Creek, West Virginia in 1912, "you will have to get it from somebody who wore the shoe, and by and by from one to the other you will get your book." This was sound advice. What follows is an effort to employ it in the case of the actual players engaged in one of the great struggles between capital and labor during the Great Depression—the Progressive Miners of Americas' strike against Peabody Coal and the UMW in central Illinois, 1932-36.

The testimonies of the miners and their families bring to life the experiences of the actual participants. Their names

Divided Kingdom

are given to show that the miners recorded are genuine historical people and their actions spontaneous, not manipulated props in a play directed entirely from above by union officers. The testimony is reproduced exactly as given so as to provide an accurate impression of the events of the time and the full flavor of these events' impact, in turn, on the miners' lives.

Included with their oral histories in the book is a selection of the photographic record as taken by amateur photographers and the Decatur *Herald and Review* cameraman in the field. Photographs documented some of the most poignant and bitter struggles of labor in the 1930's and they are used here to provide the same service on a localized scene. Thus, it is hoped a combination of their spoken word and the photographs of the scene will aid the reader in gaining a better understanding of the causes and consequences of the bitter coal strike in Central Illinois during the 1932-36 period.

The materials edited for this book were drawn from thirty-seven taped interviews conducted between April 1985 and August 1986 in and around Taylorville, Pana, Kincaid, and Gillespie, Illinois (See page 13 for a list of miners and their wives interviewed). Kevin Corley, then a teacher at Taylorville Junior High School with a Master's Degree from Sangamon State University, deserves a major share of the credit for the success of the project. Kevin conducted thirty-two of the forty interviews with miners and their wives. He also read a great deal of labor history, reviewed the tapes with me, and edited the rough drafts of the interviews. Kevin provided insights into the interviews, and took time to discuss the entire project with me over the course of four years. The collection of tapes and transcripts that we produced and deposited at the Illinois State Historical Library and in the Regional Archives at UIS should prove to be of great value to other historians and researchers.

Divided Kingdom

The Illinois State Historical Society and the Illinois Historic Preservation Agency were supportive of this project. The work was carried out with their support and they were supportive of this project. The work was carried out with the support and encouragement of Mrs. Julie Cellini, Chair of the Board of Trustees of the IHPA, and Bill Fleischli, the former Deputy Director of the IHPA. Crucial work was performed by various staff members. Janice Petterchak, then Director of Illinois State Historical Library, supported the work through the involvement of her staff in every phase. Mary Michals, the iconographer of the ISHL, assisted with the collection, reproduction, and identification of the photographs so vital in enhancing the written history. I also appreciate her trips to various sites to collect photographs. Joseph Adams, IHSL photographer, made hundreds of copies from original glass plate and film negatives for the project. I appreciate his conscientious work. The project benefitted from the word-processing work of Diane Burge, Becky DuPree, Jackie Hughes and especially Becky McCray. Ms. McCray, now a local attorney, was particularly helpful in turning copy around quickly and professionally, and offering her insightful comments. I owe my greatest intellectual and professional debt to Norine O'Brien-Davis, who typed, edited, and helped revise the final draft for production. I appreciate Norine's dedication to the project when I needed it most. The final manuscript was copy edited by Becky Bradway.

I want to thank Professor Cullom Davis and the former staff of the Oral History Office at Sangamon State University. Cullom's staff performed the absolutely essential work of transcribing and editing the project tapes for the final manuscripts. Linda Jett oversaw the entire project, which now boasts of superbly edited oral history manuscripts. During crucial periods, such as the spring of 1987

Divided Kingdom

when I was preoccupied with the ISHS History Fair, Linda performed my job of final review, and moved the manuscripts along to final proofing and typing. Elsbeth Buckley, Susan Jones, and Joyce Fisher transcribed and edited the tapes.

Other professionals supportive of the project were Dr. Ralph Stone, then of Sangamon State University, Dr. Roger Bridges, now of the Rutherford B. Hayes Presidential Center, Dr. Craig Colton, then of the Illinois State Museum, Ms. Evelyn Taylor of the Illinois Historic Preservation Agency, and Dr. John Laslett of UCLA, each of whom offered pointed criticisms of the manuscript on different occasions. The following individuals supported the project by contributing materials or donating their time for public meetings: Taylor Pensoneau of the Illinois Coal Association; Jerry Jones, then President of District 12, UMW; Professor Mark Plummer of Illinois State University; Les O'Rear of the Illinois Labor History Society; John "Doc" Davidson; and Karen Hasara.

The late Vince DeMuzio was a great supporter of all my efforts to record the history of the progressive miners in Illinois. Unknown to me, he revised his preface when he was so seriously ill the last time. I appreciate the efforts of his staff in that effort and in helping me review his revision.

I especially enjoyed the public discussions and slide show which a panel of miners and Kevin Corley conducted about the Midland Tract, 1900-1940, in the gym of Taylorville Junior High School on November 30, 1991. In preparation for that night, when four hundred community members crowded into the gym, we received support from Tom Bertucci and the staff of the Junior High School. Edna and Bob Cooper of the Taylorville Breeze-Courier supported our public publicity efforts by printing a series of lengthy articles on coal miners. Jeff Brody of the Springfield Journal-Register also published an excellent piece on coal mining in central Illinois that featured our project.

More should be said of the staff work of the Illinois Humanities Council, the public agency that funded the project. Their encouragement and patience sustained the initial project from its inception. Individual staff members who took a personal interest in our work included Dr. Jane Hood; Frank Pettis; and Dr. Robert Klaus, former director of the IHC. This particular agency is the primary supporter of grassroots humanities projects; I know of no other place in the state of Illinois where I could expect similar support.

Finally, I want to thank Bill Furry and the late Tom Teague of the Illinois State Historical Society for encouraging me to re-issue this publication. Bill contributed a great lead article in the *Illinois Times* when the book first appeared in 1991 and he still supports the kind of scholarship to which I am committed.

Carl Oblinger
Springfield, Illinois

Methodology

Until recently, historians have failed to penetrate the world of coal miners in the Midwest and Illinois. The initial problem was that there was no general work on the history of coal mining for any of the Midwestern states. In Illinois, the earliest studies of coal-mining communities focused on the ethnic composition of working populations and the wages and machinery used. During the Depression, a WPA study of coal mining communities—*Seven Stranded Coal Towns*—explored the working and living conditions of miners in seven southern and central Illinois communities, although it did not explore the impact of these conditions on the miners. Probably the most thorough study of life in coal-mining communities was Herman Lantz's *People of Coal Town*, published in 1958. The topical areas of treatment were well-chosen, including workers' attitudes, family patterns, community activities, and "social structure." Interestingly, nearly all of the information was provided by in-depth interviews Lantz conducted with the miners. Unfortunately, the work was marred by Lantz's reformist instincts and his insensitivity to changes over time.

In an attempt to penetrate the miners' lives and make sense of the interplay of family, community, and the strike of the 1930s, the two historians in this project conducted

Divided Kingdom

oral interviews with forty coal miners in Christian County and parts of Montgomery and Macoupin counties. We focused on the generation that matured in the 1920s and early 1930s, and probed our subjects' immigration experience, work, family and community lives.

In preparing for the interviews, we spent time poring over the written historical record. The criminal indictment records in the Circuit Court of Christian County, the Governor Horner Papers at the Illinois State Historical Library, and the Decatur *Herald-Review* and the *Progressive Miner* newspapers were invaluable sources. They corroborated our interviews, and linked the individual lives of the miners to broader historical actions and meanings.

What impressed me as we interviewed the coal miners was how differently two workers responded to the identical

Figure 2. The interviewing team: principal interviewer and former Taylorville Junior High School teacher Kevin Corley; Kenneth Sexson, one of the first Taylorville UMW interview subjects; Carl Oblinger, project director and author. (Courtesty of Illinois Historic Preservation Agency)

Divided Kingdom

question about an occurrence when both were present at the same event. The answer involves understanding how individual miners construct and reconstruct their memories.

In creating stories about the pivotal events of the 1930s, Midland Tract coal miners engaged in a strong tradition of storytelling in which storyteller and audience participated in the construction of a story acceptable to each. These miners talked about events, picked elements to recollect, decided how to organize and interpret these elements, and then chose a suitable narrative structure for the audience. Any narrator does this and passes it off as memory.

This reconstruction of memory occurs in the context of community, ethnic and labor groups, and social dynamics. The main question is not the accuracy of a particular memory, but why certain memories stick and are used. As far as these miners' memories go, why was there such a strong impulse to recall the family-like warmth of the community during the 1930s? Why did so many miners forget their own misery during this period? The vehemence of those remembered images was striking.

What we also saw emerging in the interviews was a deep and bitter struggle over the possession and interpretation of memories. The struggle was between the powerful imposers of change (Peabody Coal and the United Mine Workers), who sought to justify change, and the ethnic enclaves which combined private memories of a warm and unchanged past with local customs and folkways of the community and workplace. Peabody's conviction of the necessity of change for the common good could be established only by making workers forget their work traditions and community solidarity.

The finished narratives we recorded are the product of individual miners' negotiations and conflicts with others over memory. If repeated often these narratives become traditions, legends, and myths—i.e., formalized cultural expressions. For example, hoping to win sympathy for their struggles against the new industrial order in coal mining, PMA miners repeated stories about heroic events such as the Mulkeytown massacre and the battle at Mine No. 7 in January of 1933.

Criticism of oral history focuses on questions of bias, nostalgia, and representation. Since the historian of labor needs to gather information regarding the technical processes of work and the adaptation of machinery to the means of production, he will be drawn toward respondents possessing the ability to articulate the nature of technical change. This introduces a bias towards interviewing well-read and articulate people who have a strong tendency to explain the past in terms of present achievements. It is difficult to obtain an undistorted account of a respondent's observations of such events as the impact of an innovation in mining production, the PMA-UMW War, or the condition of poverty, without the taint of what was subsequently learned about their causes and consequences. Also, the person's current or recent status as a company employee or a leading trade unionist may have drawn him apart from the community or the poverty experienced by rank and file miners.

The interviews we conducted are recollections of miners at or near the end of their careers, talking about the experiences of their youth. Interviews were often tinged with nostalgia for being young and impassioned. To combat this, the interviewers could only challenge the subjects with contradictory evidence of the conditions they endured.

Finally, there remains the key issue that confronts the historian engaged in any sort of inductive methodology: to what extent can the evidence of a restricted sample be represented as the evidence of a whole population? Even with the limits of the present study—a sample restricted to a few geographically-limited core communities and to those miners still alive and willing to be interviewed—it is relatively simple to test the internal consistency of the miners' collective experience.

In comparing the unstructured autobiographical accounts of changes in the lifestyle of miners over the last fifty years, there is a consistency of experience across geographical boundaries and even wealth and ethnic lines. For example, a feeling of community runs through miners' answers, irrespective of the village, the ethnic group, or financial status. We encounter simply the communal experience of strikes and unemployment. There is also often a sense of social isolation and betrayal.

What is interesting and significant is not our interpretation of the social impact of poverty, unemployment, and war on miners and their families, but the fact that the interpretation emerges from the unreflective, firsthand accounts of our subjects. This is singularly significant in that a majority of our subjects responded so well to the in-depth autobiographical interviews.

Much of the recent promise in expanding the study of working people to include their daily lives has occurred because some historians have experimented with appropriate methodologies. The experiences of coal miners during a crucial period of transition—the 1930s—can still be reached through the recollections of a substantial number of miners and their wives. We have tried to do this in the present study.

Below, I have listed the miners and their wives whom we interviewed for the project. The tapes, transcripts, and paper work are on deposit at the Illinois State Historical Library, Illinois Historic Preservation Agency, Old State Capitol, in Springfield. (Duplicate copies are located at Sangamon State University, in Springfield, in the archives.)

The names of the miners are in order of the dates they were interviewed. Batuello and Tombazzi were interviewed by Sangamon State University's Oral History Program in 1972.

List of Miners Interviewed

John Bellaver
Tom Rosko
Kenneth Wells
Augustine Gobel
Stuart Lidster
John Sexson
Albert Morris
George Mosey
August Groh
Ada Miller
Harry McDonald
John Wittka
Merl Ahlberg
Frank Borgognoni
Frank Boch
Frederich Boch
Bob Perry
Alvin Wise
Bill Ridley
Donald Van Hooser

Joe Craggs
Perry Gilpin
Duke Allison
Frank Wingo
Edwin Olive
Otto Klein
Ola Irene Ridley
Jesse Lake
Paul "Stormy" Dixon
Kenneth Cox
Cuthbert Lambert
Sam Taylor
Lena Dougherty
Mrs. Conkey Engs
Louis Wattelet
Leo Waigausky
Emmet Bishop
Maurice Flesher
Ray Tombazzi
Jack Batuello

Divided Kingdom

Introduction

On Sunday, September 28, 1932, at 3:30 a.m., Charles H. Weineke, the sheriff of Christian County, Illinois, was informed that the Progressive Mine Workers of America (PMA) had bombed the *Daily Breeze* printing office and the offices of the sub-district of the United Mine Workers, both in Taylorville. In a deposition, for Weineke, C. F. Jewell, managing editor of the *Daily Breeze*, provided a detailed account of the events leading to the bombing and of the bombing itself: "Crowds approaching two hundred men and sympathetic to the Progressives," had since September 9 "threatened the editor and others on the streets and in the restaurants of Taylorville." On one occasion a striking Progressive, upset over Jewell's reporting of their strike, "partially tore the editor's shirt off" and called him a "rat, scab." Both the editor and his family "had been threatened and insulted by some of the leading families" in Taylorville.

The bombing of the *Breeze* printing office and insults directed at Jewell were the culmination of a series of violent attacks on the local power structure and the pretext to call in the state militia. Sheriff Weineke personally delivered to Governor Louis Emmerson a carefully

Divided Kingdom

worded, detailed communication calling for the militia. He admitted events had overtaken his feeble efforts to maintain order:

> I have faithfully endeavored to disperse the [striking] mob but it has been proved impossible for me to be at different places when needed. I have dispersed mobs and crowds in Taylorville when other mobs and crowds would be forming in different parts of the county. I have been unable to secure the public peace, and I fear I will not be able to do so in the near future.

Events would prove Sheriff Weineke's prophecy: the bombings marked the beginning of four years of violence that would shake Christian County to its foundations.

Figure 3. Bombed-out front of the Daily Breeze newspaper office in Taylorville, September 29, 1932. (Courtesy of Decatur Herald and Review)

Figure 4. Illinois National Guard officers inspect their troops in Taylorville during the mine wars, Fall, 1932. (Courtesy of Decatur Herald and Review)

Divided Kingdom

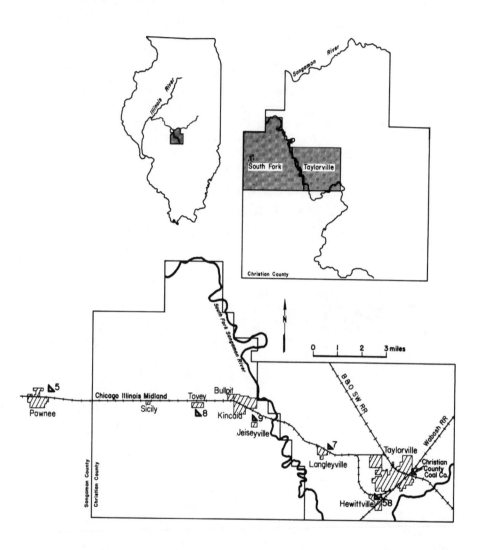

Figure 5. Location of Peabody Coal Mines and Mining towns in the Midland Tract, 1922. (Map drawn by Julianne Snide)

Divided Kingdom

I

At first glance, it seems unthinkable that coal miners in Christian County and central Illinois would challenge the United Mine Workers, a union to which they had been so closely tied by tradition and survival for over three decades. By the early 1930s, however, mining in Christian County and the rest of the state was undergoing a troublesome transition, and the UMW had embarked on a path that miners found injurious to their interests.

Beginning in the late 1910s, Peabody Coal Company, in partnership with Samuel Insull's utilities, invested substantially in the coal mines of Christian County. Peabody Coal purchased an existing company in the county and established three new mines in the small Midland Tract towns of Langleyville, Tovey, and Kincaid in 1911, 1913, and 1917 respectively (see Figure 5). At first these mines were all hand-loading operations, together employing about 3500 men. By 1932, however, Peabody Coal, with the UMW's permission, introduced loading machines, cutting machines, and conveyors into their four Christian County mines, a move that made the position of coal miners, as understood in traditional terms, obsolete. Coal miners had previously relied on their own hand techniques; the new technologies supplanted the old skills. Now, only machine operators

Divided Kingdom

were necessary to mine the coal. Also, this mechanization would throw over half of the men out of work, rearrange the entire job hierarchy, and undercut the system of seniority.

The mining communities of the Midland Tract that suffered from mechanization were separate and distinct from other communities in central Illinois. Nearly all the miners who settled these towns were immigrants and the children of immigrants. The predominate ethnic group in Tovey, Kincaid, and Langleyville was Italian. Established in 1913 by Peabody Coal, Kincaid by 1930 had a population of 1,583, two-thirds of whom, according to the U.S. Census, were Italian or had Italian parents. Tovey, a smaller town of about four hundred just to the west of Kincaid, was almost

Figure 6. Construction of Peabody Mine No. 9 in Langleyville, 1917. (Courtesy of the Boch Brothers)

Figure 7. Erecting the Chicago and Illinois Midland water tank and repair shops, 1200 block of South Cheney Street, "Hewittville," Taylorville, 1915. The C&IM operated as a coal train for Peabody Coal. (Courtesy of the Boch Brothers)

entirely Italian. Italian immigrants there had established the Tovey Miners Co-op Store in 1915. Langleyville was the last town platted in the Midland Tract (1917) and the Peabody Mine No. 7 (1918) was the last to be sunk there. The Italian miners who migrated to this town from Cherry, Illinois, lived in company-built housing.

Taylorville's mining community was different from that in the smaller towns. Most of the miners lived in the "Hewittville" section or in the north end of Taylorville and were a minority of the town's population of 7,316 in 1930 (see Figure 5). The ethnic composition in these neighborhoods was varied, with equal numbers of Italian, Slovak, British and Lithuanian among the foreign-born population. Though solidarity was more of a problem, the mining community initially shared resources and cooperated in closing the Peabody mines. As more strikebreakers were imported in late 1932 and 1933, however, fighting disturbed the fabric of the mining community.

By 1930 the miners and their families formed a distinctive ethnic and working-class society. Workers filled available jobs in soft-coal, deep-shaft mining, and were able to survive the sporadic unemployment and periodic upheavals common to the coal mining industry. The society had a strong network of kinship and friendship, expected absolute loyalty to the community, and displayed solidarity between men and women. Under normal conditions, miners drew strength from the support of their children, solicited help from family members in finding and keeping jobs, participated in collective social activities, and received emotional support from the community during periods of upheaval. But in 1932, family and community offered no barrier to the impending changes. Anxious miners turned for help to their traditional ally, the United Mine Workers (UMW), but they received no response.

Divided Kingdom

II

District 12 (Illinois) of the UMW had, since the 1890s, been a stronghold of rank and file control and democracy. Grassroots protests against unfair labor practices and the prevailing wage scale were commonplace. Immediately preceding World War I, for example, a series of wildcat strikes, supported by the United Mine Workers national leadership, protested the increased pace of mechanization and the deplorable working conditions.

Beginning in the 1920s, however, there was a great struggle over the question of district autonomy. In its early days, the UMW was a federation of separate state and local organizations which had prior histories of their own. Each was jealous of rights accorded to other organizations in the UMW constitution. In the 1920s the international office sought to enhance its authority at the expense of the districts.

Thus, the 1920s marked the beginning of an extended period of conflict between the locals, District 12, and the new international president, John L. Lewis, who collaborated with the coal companies in attempting to change the democratic structure of District 12. UMW leadership expelled twenty-four locals from District 12, and even hired gunmen to intimidate the strikers. In 1919, the international office of the UMW used a variety of means—some brutal—to force Illinois miners striking against the conditions of the Washington Agreement back to work. The international office stated that the rank and file were promoting dual unionism and attempting to destroy the UMW.

Miners were increasingly unhappy with the bureaucratic nature of the UMW. The source of the problem was John L. Lewis. In the early 1920s he consolidated his position as the chief executive official. He became dictatorial in asserting the authority of the UMW, and was less mindful of the rights of individual miners and of sub-district officials who challenged his authority. Bureaucratic growth was aided by the increasing complexity of contract negotiations and of arbitrating grievances with the Illinois Coal Association.

A number of grassroots labor movements formed in the 1920s to contest this enhanced push for UMW authority. The Progressive International Committee led the way in District 12. The committee's main objective was to preserve democratic unionism and rank and file control of the work place. To accomplish that objective the committee promoted the regulation of layoffs, joint arbitration of work rules, and, ultimately, the nationalization of mines.

Besides nationalization of the mines, the Progressive Committee and other local miners extolled the virtues of job sharing or equalization of work as a means of controlling the mechanization of the workplace and keeping operating mines in production. The notion of job sharing was very well accepted in Illinois coal fields. In 1921, for example, the Illinois Coal Association's contract with Illinois miners recognized that if a mine threw laborers out of work for thirty days, miners at another mine could, "at their option, share work with those thrown idle." The contract recognized that the practice was a time-honored tradition serving "the best interests of all working men."

By 1929, many of these ideas and time-honored practices were undermined by John L. Lewis' efforts to strengthen and

Divided Kingdom

centralize the UMW. The outcry from District 12 was so shrill that in October 1929, John L. Lewis removed District 12's charter. The deposed District 12 administration retaliated by establishing a rival body known as the Reorganized United Mine Workers of America in March 1930. After an acrimonious court fight which endorsed the UMW as the legally-recognized union, the Reorganized Miners disbanded.

The miners' frustration at the economic crisis in the coal industry and the inability of the UMW of A to protect them led to their willingness to join splinter groups and defend traditional practices. The central problem was the high cost of mining Illinois coal, which could not compete with the cheaper coal that was flooding the market by 1930. In addition, the Jacksonville pay scale established in 1923 at $7.50 a day for labor boosted wages and priced Illinois coal out of the market.

To meet this crisis, John L. Lewis decided by the early 1930s to work with the coal companies to reduce wages, reduce the labor force and mechanize. As Harry McDonald, an engineer at Peabody's Mine No. 58, explained in an interview in 1986:

> **J. L. Lewis saw the wave of the future. Mines would shut down because of the excessive [labor] costs. He felt no one could make it under the old ways because they'd still do it by hand. If they had to let the men go, then they would have to do it. That's all. You can't go under some old bent up way; you have to go with progress.**

How far the union had moved from its membership was demonstrated in the 1932 contract talks between the UMW and the coal operators. Every Illinois local of the UMW, including the four in Christian County, demanded a substantial reduction in the length of the work week, improved grievance procedures, guaranteed seniority, a limited phasing in of the new machinery, and the sharing of available work under the supervision of the local UMW president.

Divided Kingdom

District 12 President John Walker presented these demands at the first negotiating session with the Illinois Coal Operators' Association. The operators rejected their demands, proposing instead a thirty percent wage cut; no change in hours and overtime; a "streamlined" grievance procedure; ways of speeding up the introduction of machinery into the mines; and the concentration of production at certain "efficient" mines.

Seeing no possibility of compromise, Illinois miners of District 12 went on strike on April 1, 1932. On July 16 the rank and file rejected two-to-one a settlement offered by UMW leaders with the same provisions as previously proposed by the Illinois Coal Operators' Association. At that point Walker called in John L. Lewis to negotiate a settlement. On August 10 Lewis promptly rammed the same contract through ("stole the election," according to the miners) against the overwhelming wishes of the rank and file and ordered Illinois miners back to work on an emergency basis.

But it would take more than Lewis's order to force the miners back to work. Rank and file miners organized mass picket lines to halt the operation of mines reopened under the terms of the new contract. In August 1932 picket lines succeeded in closing all operations in Randolph, St. Clair, Madison, Christian, Washington, Sangamon, and Fulton counties. Over the protest of the rank and file, John L. Lewis brought in strikebreakers from other districts to operate the mines under the new contract. That action completed Illinois miners' disillusionment with the UMW, and on September 1, 1932, a disgruntled group of UMW rank and file organized the Progressive Miners of America in Gillespie, Illinois.

In the initial months, the PMA secured a broad base of support in the central and southern Illinois coal fields with an estimated 20,000 members. The PMA attracted second generation Italian and Slavic miners who had not previously been involved in UMW affairs. One of the key planks of the new union's platform was opposition to anti-foreign sentiment, which had prevented effective ethnic participation in the UMW. Many ethnic miners felt that John L. Lewis' blatant disregard of the democracy of the local had to be checked by the PMA.

The PMA was committed to the ideal of democratic trade unionism and the older traditions of District 12, UMW. No miner could be elected to office in the PMA without having worked in the mines for at least five years. Union organizers were appointed for a period of no more than four years and were not permitted to vote at union conventions. Provision was also made for the recall of elected officers and the calling of special conventions at the initiative of local unions.

The PMA scored initial successes throughout Illinois. In the fall of 1932, many mines outside Christian County reopened under a new PMA contract that granted employment to the greatest number of men possible and maintained the same $5.00 scale offered under the UMW's contract. In a 1972 interview with Sangamon State University, Joe Ozanic, who in 1932 was the local's president in Mt. Olive, recalled:

> The local [PMA] just simply adopted the rule that no man is going to work overtime without showing cause why no one else was available to share the work. We enforced this in order to equalize employment and got the company to train all to operate the machines. We got the company to go along by getting the key men in our union to cut down the production. Men's lives, and bread and butter for the kids, are just as important as the production profits for coal companies. The better companies grudgingly accepted this.

Peabody Coal, the largest and most advanced of the coal companies in Illinois, opposed reopening under the PMA contract and made extraordinary efforts to defeat the new union and the striking miners. Peabody insisted on reopening its four Christian County mines under the discredited UMW contract. They hired large numbers of

Divided Kingdom

strikebreakers from southern Illinois; imported gun-toting "enforcers" from the same southern Illinois coalfields; and organized an extensive campaign of sabotage aimed at discrediting the striking miners.

Peabody's campaign paid dividends: Mine No. 58 in Taylorville reopened in late September 1932. Gradually manpower was sufficient to reopen the other three mines by January 1933. It was estimated that the labor force had been reduced by half when all four mines were again functioning, and that more than half of that labor force were strikebreakers from southern Illinois and Kentucky. Some miners pointed out that Peabody did not want to settle, since they would have to take back the striking miners.

Peabody Coal's production policies paralleled its plans to defeat the striking miners. The company concentrated production at its mines in Christian County, where they were mechanizing more efficiently, and closed operations elsewhere in central and southern Illinois. These practices created massive unemployment in Franklin and Williamson counties, irregular employment in Peabody's Springfield mines, and full-time employment for a limited number in Christian County.

Divided Kingdom

III

Witnesses to the bombings, beatings, shootings and violence that ravaged the Midland Tract between 1932 and 1936 believed the turmoil was caused by rival unions fighting for power. While a few PMA leaders saw the UMW as simply a competitor for influence, striking miners viewed the old union of John L. Lewis as corrupt and an obstacle to steady employment.

Most striking miners thought that the corruption of the UMW and the miners' unemployment could be traced to a conspiracy between union officials and Peabody Coal. In Christian and Sangamon counties, for example, only Peabody defied striking miners by refusing to allow a vote for union representation or by honoring the election for representation conducted by the National Labor Relations Board. Time and time again in 1932, Peabody Coal refused official requests to hold representative elections at their mines in West Frankfort and the Taylorville area. At Mine B in Springfield, for example, the 1937 representation vote showed 425 for PMA and 24 for UMW, but Peabody kept the mine closed until a favorable vote was forced. Art Gramlich, a retired PMA member in Springfield, recalled in a 1979 Sangamon State University interview: "We discovered that the UMW was paying the

Divided Kingdom

Figure 8. Miners on strike. Massive picket line outside Peabody Mine No. 58, August 29, 1932. Peabody attempted to bring in strikebreakers, seen here in the vehicle stalled in the middle of the picture. (Courtesy of *Decatur Herald and Review*)

Figure 9. Pickets keep Peabody Mine No. 9 in Langleyville closed, August 23, 1932. Without National Guard troops the pickets intimidated strike-breaking miners. (Courtesy of *Decatur Herald and Review*)

Divided Kingdom

damned *mine owners* to keep their mines closed while we nearly starved to death."

What the striking miners resented most, though, was the UMW bringing in miners and "intimidators" from southern Illinois mines—mostly Peabody mines—to cross the picket lines to operate Peabody mines in Christian and Sangamon counties. "It beat all hell," remembered Joe Ozanic, "to see the militia escort scabs from all over the United States who worked for wage of five dollars, UMW scale, and who were paid for and provided by John L. Lewis. Hell, John L. even got thugs on his payroll and assigned them to Christian County to hurt us."

The most important feature of the striking miners' protest was its collective nature. Entire communities, such as Langleyville, Tovey, and Kincaid, "almost to the man," picketed, protested, and intimidated men who attempted to mine coal at Peabody Mines No. 58, 7, 8, and 9. "You could not believe how it was," related John Pendias, a company man at No. 58, in a sworn deposition taken in 1933: "I received a pounding from a gang of strikers on September 3 in front of Harbarger's Garage in Stonington. No one dared raise a finger in my defense." Sheriff Weineke, once full of bravado, said that "crowds of striking coal miners" which assembled day and night in Langleyville, Tovey, Sicily, and Kincaid "have created a state of lawlessness. These places, to a man, are incensed at the bringing in of strikebreakers."

The women and children of striking families also participated in all phases of the communal protest. Cuthbert Lambert, who rejoined the UMW in 1934 after two years of striking, remembered that "the whole family belonged to the Progressive Union, went to the meetings, went on the picket lines, and worked. See, no one could protect us, and the government didn't give us nothing. We took care of our own—women, men, and children."

Divided Kingdom

Women also were present in the strike activities. Thousands of women joined the PMA Women's Auxiliary. Led by spokeswomen such as Agnes Wieck of Belleville, Ruth Besson of Tovey, and Celina Burrell of Langleyville, the PMA Women's Auxiliary pushed their men into breaking with the UMW and organized huge rallies. Sometimes they confronted the coal bosses themselves. In the fall of 1933, more than ten thousand Auxiliary women and their families attended a series of rallies in and around Taylorville that denounced the new National Recovery Act Code and the UMW. (The NRA legitimized the exclusive relationship of the UMW with the coal operators, and excluded the PMA as one of the recognized agents for the coal miners.) In the fall of 1936, when thirty-eight PMA miners were tried and convicted of dynamiting the Chicago and Illinois Midland tracks in Christian County and company property elsewhere, ten thousand Auxiliary women staged a sit-down strike at the Governor's Mansion in Springfield.

Women were involved in the most serious confrontations of the picketing. Throughout the 1932-1933 violence at the Peabody mines in Christian County, Auxiliary women accosted the scabs in front of the mines. Forty years later, Joe Ozanic told the story with pride: "...and they weren't afraid of the hired thugs and militia. They'd say to the scabs, 'What are you doing here? These men are on strike against a big company union and corruption.' To the militia, 'What are you doing carrying a rifle in the militia? Aren't you ashamed of yourself?'"

As trouble escalated in the Christian County coal fields, the support of the Auxiliary was essential. Women walked the picket lines and collected food from farmers and distributed it to local commissaries where PMA miners were on strike. In Christian County, the Auxiliary ran three commissaries for the relief of striking miners and their families. On the state level, Auxiliary leaders formulated the key

Divided Kingdom

PMA policies that kept the union cohesive during the organization's first two years.

Sharing, cooperation, and collective decision-making for the common good were central to the life of the Midland Tract communities. It was no accident that the PMA protest, emerging as it did from the ranks of the mining communities, involved entire families and communities. This protest ultimately sought a sharing of available work, collective decision-making, and modification of the pace and the way machinery was being introduced into the mines.

Nowhere were the concepts of sharing and cooperation better learned than in the daily routine of a miner's family life. Childhood was followed by an early departure from school and entry into the mines. A child's earnings were regularly turned over to parents and kin, and this support was given unconditionally at least until marriage, and sometimes after. "There was no other way," confessed Alvin Wise, a miner from Stonington, "for every bit counted in the struggle to survive."

The family obligations that dictated wage sharing and an early entry into the mines were reinforced by dependence on kin and community for work. Since a miner could choose his helper and influence the selection of slate pickers, he had a major influence in the hiring of labor. In almost every case, he chose relatives or members of his own ethnic group. During a strike, a miner could still find work for his kin through relatives in Detroit, Chicago, and other industrial communities.

The community beyond the family was also important for survival. Striking miners in Christian County could not depend on a reliable income, especially since the cash economy had evaporated, so a flourishing barter economy and a strong help ethic developed. "My dad needed to dig his basement," related Frank Borgognoni, a miner from Kincaid. "No one said anything. They just come and pitched in.

Divided Kingdom

The women didn't even ask the men where they were going. They put together food and prepared a good meal for the men." Groups of family members and neighbors prepared smoked sausage and sauerkraut, and canned fruits and vegetables. Neighbors shared food with those temporarily in short supply and prepared neighborhood suppers for festive occasions. Many Christian County residents picked coal along the railroad tracks for the elderly and disabled.

Communal sharing and helping became a means of survival in times of strikes. An investigation of the 1932 strike by the Adjutant General's office reached some striking conclusions: "Rank and file miners have a long experience with hardship. They can resist the coal company since they share their meager resources." In particular, Peabody Coal's eviction of striking miners from the homes they rented "was useless. They double up with other families, and draw sympathetic comment on their plight."

Non-striking PMA miners in Montgomery and Macoupin county coal mines donated as much as a day's wage to the Christian County PMA strike fund and offered home grown food to feed the out-of-work miners. Joe Ozanic's description is poignant:

> In Christian County in the winter of 1932-33 Peabody had five [sic] mines out of work. Voluntary contributions of working PMA members helped striking brothers and their families. They even built commissaries that could feed those that couldn't feed themselves, and those with homes provided homes for those ousted from company homes. That was the background of our movement—how we managed to fight, to have the spirit to offset what we lacked otherwise.

Miners and families who deviated from communal goals—particularly strikebreakers—were threatened, beaten, or shunned. "I have not yet decided to cooperate with the strikers and agree not to work," testified Paul Clayton, a yardman at Peabody No. 58, in Taylorville, on September 18,

Divided Kingdom

Figure 10. Community solidarity, October 23, 1932. PMA Auxiliary women, in their white hats, wait in line to view the casket of a miner's wife killed in Tovey. (Courtesy of *Decatur Herald and Review*)

Figure 11. Communities organized soup kitchens to feed the striking miners and their families. This "distribution station" in Taylorville, organized in September, 1932, dispensed groceries and bread. (Courtesy of *Decatur Herald and Review*)

1932. "Only last night I took my wife to St. Louis . . . because we have received threats that our house was to be blowed up. I left, and have not made up my mind where I will live or with whom I will stay." John Wittka had the same experience in Langleyville; but in this case the striking miners not only threatened but administered a severe beating in November 1932. "I would not, *never, ever*, go through anything like that again," related John Wittka in an interview in 1986.

Striking miners made a special communal effort to censure and isolate scabs and their sympathizers. There certainly were limits to this cooperation, but it is hard evidence to the contrary. The massive number of striking miners and their wives on picket lines, the functioning of communal kitchens, the extreme level of rhetoric, and even the strike of high school students in Kincaid because the school used Peabody coal, all point to a strongly cohesive Christian County mining community in the early 1930s.

Divided Kingdom

IV

The PMA suffered a reversal of fortune in the late 1930s. Most mines under contract with the Progressives in 1932 were small and could be mechanized only with great difficulty. Consequently, they could not compete with the larger mechanized mines under contract with the UMW; more than half of the mines under contract with the PMA no longer existed in 1939. From more than twenty thousand rank and file members in 1932, PMA numbers dwindled to less than five thousand in 1940.

UMW-backed Peabody Coal, although able to reopen only Mine No. 9 in 1932, had four Peabody Coal mines producing at full capacity by the end of 1932. Even more important, the mechanization of those mines was being completed and output more than doubled during the same two years. Employment went from zero to nearly two thousand by 1935.

The PMA suffered from the increased efficiency of the mines under contract with the UMW. Mechanization of the Peabody mines and other coal companies' properties under UMW contract reduced the resources upon which the striking miners could draw. The introduction of the loader and conveyor belts increased production at mines, allowing them to reduce coal prices, hire more labor, and squeeze

Divided Kingdom

existing PMA mines still further. Striking miners in Christian County could not remain unemployed indefinitely; by the mid-1930s their children were undernourished, many had lost their homes, and the PMA commissaries had closed their doors. Eventually, the miners were confronted with two choices: rejoin the UMW and work in the Peabody mines, or quit the Illinois coal fields for the auto factories of Detroit or jobs in Chicago and St. Louis.

The non-competitive nature of the PMA mines and the advantage mechanization gave to Peabody Coal were decisive factors in the demise of the PMA and the capitulation of the striking miners. But the most chilling factor was the 1936 indictment of forty-one PMA miners and strikers for bombing the Chicago and Illinois Midland Railroad tracks near Taylorville. Thirty-eight of the miners—six from Christian County—were convicted in 1938 and served time in Joliet and Leavenworth prisons. That action, many in Christian County believed, broke the spirit of the strikers and strengthened the hand of the UMW. After that episode, peace existed in the Christian County coal fields for the first time since 1932.

Divided Kingdom

A Chronology of Peabody Coal's Establishment of Mines and Towns in Christian County

1900 Springfield District Coal Company sunk new mine in Hewittville.

1901 Christian County Coal Company opened mine between East Main and East Market Streets, Taylorville. Closed in 1908.

1904 Christian County Coal Company started mining coal in Stonington. Mine sold to F.S. Peabody in 1916 and became known as Peabody Mine No. 21 until it closed in 1924.

1911 Peabody Mine No. 7 sunk in September at site of Kincaid. Peabody Mine No. 8 sunk same month at site of Tovey.

1913 Kincaid founded by F.S. Peabody. Under supervision of the Kincaid Land Association—a subsidiary of Peabody—workers built five businesses, eighty-five homes and the town's infrastructure.

Divided Kingdom

1915 Village of Kincaid incorporated, January. Tovey founded by miners. Tovey Co-operative store opened, sponsored by Italian immigrants. Bulpitt incorporated.

1917 Langley family platted Langley. All coal and mineral rights sold to Peabody Coal Company.
Peabody sunk Mine No. 9 at Langleyville in the fall.

1918 First coal at Langleyville hoisted June 21.

1921 Peabody Coal company purchased the Springfield District Mine, known as Peabody Mine 58.

1951 Peabody Coal Mine No. 9 ceased operations March 31. Peabody Coal Mine No. 10 began operations June 13. Claimed to be the largest underground mine in the world.

1952 Peabody Coal Mine No. 7 ceased operations May 29.

1954 Peabody Coal Mine No. 8 ceased operations July 27.

Divided Kingdom

Figure 12. Boch family in Hewittville, early 1900s. (Courtesy of the Boch Brothers)

Chapter I
Family and Community: Sinking New Roots

My mother was sick a lot, see. She had five major operations in her lifetime and two nervous breakdowns. If you had to live like we did then, you'd understand.
—Frank Borgognoni

Intricately woven kinship associations ran through the fabric of coal-mining life in Christian County during the first three decades of the twentieth century. At every turn in life, relatives and family stood ready to assist one another. Whether one needed a place to live, a job, money or emotional support, family was involved. Such relationships were based primarily on reciprocity between parents and children, husbands and wives, the immediate family, and more distant relatives.

Beginning in 1911, certain neighborhoods in Taylorville and the coal towns of the Midland Tract were settled by southern and eastern European immigrants who secured jobs through relatives who already worked for Peabody Coal. Newly arriving male immigrants almost always had previous mining experience and when jobs became available they easily filled the vacancies. Sons followed fathers and uncles into the mines, some starting as young as thirteen.

The interviews with Joe Craggs, George Mosey, and Stuart Lidster found in this chapter detail the experiences of British-born first-generation coal miners who were attracted to the Peabody mines between 1912-1922. These miners brought knowledge of long-wall mining methods similar to those methods used in Christian County; provided help for other newly arriving immigrants; brought important work traditions and notions of solidarity; and established positions of authority in the work force. Their longer prior experience was, therefore, a crucial link in socializing new immigrants to industrial culture. The majority of these Anglo-Saxon immigrants eventually found employment in the mines as superintendents or managers, unlike the southern and eastern European newcomer who was most often consigned to menial hand-loading operations and rarely rose to management.

Ethnic miners from southern and eastern European backgrounds drew upon the traditions and habits acquired in pre-industrial families to sustain themselves. For example, miners found agricultural talents very useful, as demonstrated in the Frank Boch and John Sexson interviews. Gardening and livestock raising were widespread, and nearly all planted gardens and kept cows, pigs, or chickens. Gardening was not simply an attempt to cling to a previous way of life. Rather, it was a highly important economic safety valve in an industry plagued with irregular employment and periodic depression. Gardens saved the miners during work stoppages and provided fresh produce for the dinner table and work for the unemployed.

The interviews with Frank Borgognoni, Lena Dougherty, Tom Rosko, August Groh, and Ada Miller emphasize other crucial functions families assumed for southern and eastern European immigrants. The interviews show how precarious

Divided Kingdom

life was: men as breadwinners, died young. Italian and other eastern European immigrants, more so than the native-born and the British, clustered together for mutual support and moved about between ethnic enclaves looking for jobs. Ethnic miners seldom came directly to Christian County. Many came to this country and settled initially in the northern Illinois coal fields near Ladd and Spring Valley, and most eventually tried to return to the old country after they had saved enough money.

Once established in the community, the family unit continued to serve important and necessary functions. In addition to serving as sources of job information for a shifting, fluid work force, families were a strong support system during periods of emotional stress, old age, death, strikes, and shut downs. Children were essential breadwinners during hard times, especially when heads of households were injured or died (a common occurrence in the Christian County coal mines). Many of our interview subjects left school early to assist immediate and extended families. Children held an obvious respect for parental authority. They did not question the actions of their parents and were expected to obey them.

The Frank Borgognoni interview demonstrates the family's link with the community. The community often acted as surrogate parents, sharing food and work with those who resided nearby and were in need of assistance. The community carried on functions that the family could not provide, such as pitching in to complete major projects, caring for the sick and elderly, sharing job information, and carrying miners on the "book" in tough times.

Communal drinking was a major hallmark of mining town life, a means of solidarity and of supplementing income, as shown in the Dougherty and Borgognoni interviews. Miners drank to celebrate births, christenings, weddings, holidays, and paydays. They made their own liquor, using almost any-

Divided Kingdom

Figure 13. Young rock pickers from Peabody Mine No. 58 enjoy an outing with their boss at Kimlake, Christian County, 1926. The boys are drinking home brew. (Courtesy of the Boch Brothers)

thing in it, including the crops they grew. Each ethnic group made its own favorites. Most specialized in homemade beer, called "home brew." Italian miners were fond of making wine and a potion called "pickhandle" from raisins, yeast, and water. The making of homemade liquor was not always for their own consumption. During Prohibition much was manufactured for bootlegging or reselling to other miners and even outsiders at cheaper rates than could be had at commercial establishments.

The success of moonshining depended upon community cooperation. The community provided the labor necessary for production and the distribution networks. Community-wide discretion was essential for success of any venture.

The prominent taverns in coal towns were run by retired miners or ethnic compatriots and served a social, as well as an economic, purpose. Since the taverns' clientele was so diverse and so polyglot, a welcome was extended to every mine worker regardless of ethnic origin, creed, or religion.

The basis for community life came from a heterogeneous mixture of nationalities and native-born such as in Taylorville and Stonington. The percentage of mixed foreign-born in the two towns' population grew from less than five percent in 1900 to over a third by 1930. Other communities were isolated patch towns where one or two ethnic groups were predominant, such as in Tovey, where Italian miners comprised ninety percent of the population; and Bulpitt, where Lithuanians dominated. These two towns, as well as Langleyville and Kincaid, did not even exist in 1910.

No matter what the ethnic mix, the coal towns—especially Kincaid, Langleyville, Tovey, and Bulpitt, and, to a lesser degree, the north end of Taylorville—were alienated from the social and political connections of the larger society. Sheriff's and State's Attorney's officers regularly extorted money or arrested hundreds of ethnic miners on charges ranging from bootlegging to "civil disturbances." Most of the charges had no substance and were filed to put pressure on the miners to remain employed with Peabody and to keep "civil order." Recognition of uniqueness, isolation, and impotence fostered community among miners. Loyalty to each other was based on the understanding that theirs was a separate society and that economic security and basic civil rights were not available to them under the current system.

By 1920 there was strong reliance on fellow miners and a community network that offered information on jobs, provided resources, and even supplied the work force. Though harassed, miners were the most fiercely independent of industrial workers: they made their own decisions and enforced ethnic, community, and occupational cohesion. There was a common community of labor in all of the downstate mine fields because they had bound themselves to coal mining as a way of life.

As the skills of the miners were diluted in the 1920s, the community became more important. Miners found sustenance and satisfaction in their families and communities. Preoccupation with the concerns of day-to-day life, family, work, and community celebrations spelled economic and cultural survival.

Divided Kingdom

Figure 14. Taylorville's first large mine, the Christian County Coal Mine, was located in the 800 block of East Franklin between East Main and East Market. The mine closed operations in 1908. (Courtesy of the Boch Brothers)

Figure 15. Miners, nearly all of British stock, in front of the Christian County Coal Mine in the early 1900s. (Courtesy of the Boch Brothers)

Divided Kingdom

British Immigrants

Sam Taylor

Born in Taylorville in 1910, Sam Taylor lived his entire life in Christian County and was an excellent observer of the mining wars and local conditions. As a third generation lawyer, Mr. Taylor took special interest in the settlement of Christian County's coal mining towns, and the land ownership claims of Peabody Coal.

In the north end of Taylorville where I lived was a Latter Day Saints church. Many of the old time coal miners had come from England where they had been converted to the Mormon religion and then had been solicited for the original coal mines in Taylorville before Peabody bought 58 and laid out Kincaid. The old mine was east of town and it had been abandoned by the time 58 was built. The 58 mine was originally started by local people who had started the old mine and was bought by Peabody. Then Peabody sank the three mines out in the Midland and laid out the town of Kincaid.

The town of Kincaid was laid out by Peabody Coal Company, plus Commonwealth Edison bought around a thousand acres of land for the coal mines to lay out this town. They bought the land from half-way to China and when they sold lots they sold only the surface rights. So the coal mines that are operating right now and the village of Kincaid were a hundred percent Peabody Coal Company, and the people who lived in Kincaid never got a penny out of it.

Stuart Lidster

Stuart Lidster's father was not the typical British coal mining immigrant. Unlike the majority of skilled Anglo immigrants who migrated to the U.S. coal fields, Lidster was an unskilled laborer who arrived after the main body of British laborers in the late nineteenth century.

I was born September 8, 1907, in London. My dad was in Nokomis by 1912. He worked at the North Mine in Nokomis and then he went back. My mother died in 1918. In 1921, we came over here again and my dad worked sixty-three days. 1921 was a bad year.

I wanted to stay with my grandmother. I didn't want to come to Illinois, but I got adjusted. It didn't take me long to learn the difference between fifty cents and half a crown—that was English money.

I was all over when I was on that ship to the U. S. We was third class, of course. Some other passengers give me maybe fifty cents. I knew fifty cents was worth more than half a crown, so I'd have an extra half a crown in my pocket. I'd give the steward the half a crown for three oranges. I didn't get seasick until the fourth or fifth day. We didn't have anything to do, just skipped rope and stuff on the deck, you know. All different kinds of nationalities sailed, of course. We sailed from South Hampton, stopped at Cherbourg [France] and then came across. I guess a lot of Europeans got on at Cherbourg.

Divided Kingdom

I had a job picking strawberries two weeks after I hit Nokomis. My dad died in August and I started picking rock September the seventh, the day before my sixteenth birthday. But you got to bear in mind that the mine only worked about two days a week. If I wanted a day's work, I had to start before my birthday. You're supposed to be sixteen but I wasn't.

We came because we had two sisters in Nokomis. They came eight years before my dad did. I'm sure my one aunt came in 1907. Her husband worked in a mine in Nottinghshire, England. I think he started working in Kentucky in the mountains, in the mine. Then he moved to Athens, Illinois.

Joe Craggs

Joe Craggs is more typical of the skilled British laborers who migrated to the Midland Tract. The family started mining in northern Illinois and moved to central Illinois when the better fields were developed after 1910.

The miners in Christian County came mostly from England, Ireland; some Scots, Italian, Croatian, and Hungarian—those countries in Europe that had coal mines. They were the second generation coal miners. There was a few first generation coal miners in my time, and that is what made the world great. It didn't make any difference whether he was a Frenchman or an Englishman or an Italian. They were comrades in their work.

I can't say that there was a language problem because the common bond was that they knew what their job was—to mine coal—and they were easily taught to mine coal. The majority of them. Some were good for one job and some were good for another job. So, you had the Scotch and English that were real good timber people. At that time we didn't use roof bolts; roofs bolts were a completely different story. We used

Divided Kingdom

timber and we used men that were experts at it. That knowledge all came from long-wall mining in Great Britain and other countries over there. It was brought into the United States.

I was born September the second, 1916, in St. David, Illinois, a little mining town in the northern part of Illinois. My family came from England to a little town called Norris, Illinois, which is about twelve miles from St. David. My father was a coal miner in England where he started working when he was eleven years old. He and my mother were married in the old country and they came to this country in 1909 or something like that, and he worked in the coal mines in Fulton County for a great number of years.

Let me tell you more about his early experiences in the mines in northern Illinois. When I was a little boy—a real little boy—he started his own coal mine, at that time we called them country banks. Up there the coal crops out in the side of the hills. In the summertime lots of coal mines shut down because of lack of business. So maybe two or three miners would get together and would go into the countryside and make a deal with the farmers and start what they called a country bank—their own little coal mine in the side of the hill. They would mine the coal and get them a stockpile so in the wintertime they would have coal to sell. So they would have a living that way. They didn't make much in the summertime, but they got their hole started in the ground. They would work that until they either worked it out or it became no longer economical. My father was superintendent over several of those. I used to go there when I was a kid and mess around when I was just a little tiny shaver.

One time my dad came to Springfield—to the State Fair—and he met a man by the name of Harry Thornton, who he knew in England. And Thornton talked him into coming to Taylorville and working for Peabody Coal Company.

The Thorntons, who got Dad to come, ran a hotel across from mine No. 8's tipple in Kincaid. It was a big, two-story

Divided Kingdom

hotel where all the coal miners boarded from mine No. 8. In those days the coal miners were coming from the old country, and from Italy, all over. They didn't have any place to go. They would come to the Thorntons. The Thorntons were good English people. When I was a kid we used to take hot lunches across the street at noon for the men to eat at the boarding house.

Harry Thornton's son-in-law was a man by the name of Johnny Hardy. He was a mine manager, a superintendent in mine No. 9. When he died he was the operator's commissioner to settle labor cases. That's how my dad got started in this country mining with Peabody Coal Company.

Years ago, you almost had to be an Englishman or a Johnny Bull to be a foreman. There was very few other nationalities that were foremans because they came to this country with more experience because the English laws in coal mining had developed much further as far as safety lamps. The first safety lamp was developed by an Englishman named Davey. They knew gases and what have you, and timbering; they were experts at that.

George W. Mosey

Many British immigrants settled where relatives were working. Most British-born workers generally started work in positions above common laborer; Mosey was a trip rider, a mine laborer who rode on the front of the small train that hauled coal out of the mine.*

I was born in Sunnyside, England on November 28, 1891, and I lived there until I was twenty-two years old.

My father was a coal miner. I started coal mining at thirteen. Where my father worked, they wouldn't hire anybody until they were fourteen, but there was a company that would hire you under age about three miles from where I lived, in a place by the name of Catchgate. So I worked in that mine, and the feature about that time—it didn't have no fan. It had an air shaft, but they had a fire in it to circulate the air. So I worked there until I was fourteen, and then I went and worked where my father was at. I drove a pony until I was sixteen years old.

I had to drive a pony until I was sixteen, then you could go and do what they called a putter. That's taking the cars to the diggers. Until you were twenty-one; then you could go to the face* and load coal. So I loaded coal until I was twenty-two.

I had an uncle living in Virden by the name of William Shears. I don't know how long he'd been in this country, but he sent me a ticket to come. Do you know the price of a ticket from Liverpool to Virden? 58 dollars. That included a ship and the railroad fare. Well, I landed in Boston and the ship I come over with was the *Arabic*.

In Boston I got on the B & O to Chicago. Then I took the Chicago Northern. That was a railroad that ran from Chicago to St. Louis. I got on at Chicago and I was hungry. I finally got something to eat on the train. Of course, I had to pay for it. I could read the signs passing the station. It stopped at Springfield and Litchfield I think, this train. So I kept looking at the signs, and I see a sign, Virden. That's where I wanted to get off, and the train kept going. So I told the porter, "That's where I'm supposed to get off at. So he pulled the rope about a mile down the track. There I had two suitcases, walking down the track to the depot. So I get to the depot and there happened to be a relation of mine working at the depot.

So I went to their house and ate my first meal. Their name was Rankin. They took me to where I was going to live, to the Shears' place, with my uncle. I worked about a year at what they called the West Mine in Virden.

Then I moved to Tovey, which was building up at that time. That would be about 1914. And I worked at No. 8 for fourteen years. I started low down like every single company man used to have to do. I'd couple cars* and ride trips*. I used to ride for different fellows, you know. I worked trip ride for a fellow by the name of John Hardy. He used to be superintendent at No. 9.

For all terms marked '' see the glossary, page 255.

Southern and Eastern European Immigrants

August Groh

August Groh's family was one of the first eastern European families to migrate to the Taylorville area to work in Peabody's new mines in Pawnee around the turn of the century. Word of the new mines was passed by relatives through the entire family network.

I was born August 26, 1898. I was only about five years old when my folks left Europe. I was born in Europe—Austria-Hungary. I was born at a small village. It sounds French—Rashichia. R-A-S-H-I-C-H-I-A, I think it is.

My dad came to this country a year before we did. After he was here a year he sent for us and we came over on a ship named the *Hamburg*. We departed from Hamburg, Germany, and we landed at Ellis Island. Of course, we talked German, see. And at that time, the porters all knew German. I had a sister older than I. These porters took care of the three of us and showed us where to go. We couldn't talk English and they routed us to Chicago. From there we went on to Lincoln where we stayed about six months. My dad worked in a mine at Lincoln and he heard that Peabody was going to

Divided Kingdom

start a mine at Pawnee, Illinois. So he decided to try and get work down here. He was lucky enough—he got a job—and that's when we came here, somewhere in late 1904.

We bought a little house and my dad kept working in the mines. We moved over to this place from Lincoln in about 1905. When I was sixteen years old my dad thought maybe he'd try to get me to working. At that time I was a trapper*. A trapper takes care of the main haulage lines* where they pour the coal. I opened the door for them to get through, and after they were through, made sure the door was closed due to the fact that a door is there to circulate the air. If I left that door open, then the men inside would be deprived of that much air. So I had to keep the doors shut—when they weren't going through it. I did that for, I'd say, a year or so, and I got big after that so that I could load coal. So my dad got a room*—that's a term for the working place for the miner. So I stayed there, worked for my dad I would say four or five years, hand loading coal.

Frank Boch

The Boch family migrated from Austria around the turn of the century. The elder Boch came in the early 1900s to work first in Springfield, and then in the old Taylorville Mine before Peabody bought it out.

I was born in Taylorville, Illinois, right here in his house, October 6, 1908. There was a mine here and houses were built around the mine. It was known as Hewittville because at one time L.D. Hewitt, a banker in Taylorville, owned all the property here, or a lot of it.

Our folks come from Austria. My parents didn't come over to this country at the same time. My dad come in 1896 and my mother didn't come till about 1907. They already

Divided Kingdom

Figure 16. Photograph taken from the top of Stephen Boch's house in Hewittville, 1929. In the background, left, is the old Hewitt School; in the middle are the Chicago and Illinois Midland repair shops. The mine, which became Peabody Mine No. 58, was sunk about 1898. Hewittville developed around No. 58 as an ethnic miners' community. (Courtesy of the Boch Brothers)

Figure 17. Peabody Mine No. 58 in the early 1930s. Known originally as the Springfield District Mine Co., it was located on Houston Street in Hewittville. (Courtesy of the Boch Brothers)

Divided Kingdom

had two children in Austria. I'm the oldest one that was born in this country and there was an age gap of about fourteen years between the children that were born there and the ones born here.

My dad come to Springfield first, I believe, and worked some in the mine, but it was a bad time. They didn't call it a depression like we do now; they called it a panic and there was no work. He joined the United States Army and was a veteran of the Spanish-American War. He went down to Cuba and Puerto Rico and those places where they were fighting. After the Spanish-American War, he came to Taylorville and bought this property here. He bought one acre here—these six lots—and there was a whole block over there. Of course, I don't know what he paid for it, but he couldn't have paid much, because I don't think he had much money. It was probably cheap at that time. He saved some money from the service. Anyway, that is how we became established here. Then the old mine was sunk about 1898 and he got a job at the mine.

Oh, he used to tell a story about the old mine. He was always full of jokes and he was wanting to get on at the mine. They wouldn't hire him; they didn't have no room for him. One day one miner got killed at the mine and the mine manager, or superintendent, lived a short ways down here, from our house about a couple of blocks. They come after him and said, "Steve, come on to work; we got a job for you. A man got killed in the mine yesterday." And my dad said, "Well, if I had known that's all it took to get a job, I would have killed him a long time ago." He was always full of stuff like that.

He started in the mine here in Taylorville about 1900. It was brand new. It was just sunk by the Springfield District Coal Mine Company. There was business people here from Taylorville that put their money into it and started a mine. In 1921 or so Peabody bought the mine and then it was known as Peabody Mine 58 from then on. In 1924, when I started, Peabody had this mine already.

My dad worked in the mines in Germany and he was surprised at how they mine here. He says in Germany they would go in with their hands and scrape every bit of the coal that they could get, you know, and put it on their shovels and put it in the cars to take out. Here in the United Stated, I have seen it because I have worked here—why, tons and tons of coal is left down. They get what is easier to get, you know, and they don't bother about what is hard to get. He says there, they get every bit of coal, every pound that they can get; they don't waste none of it. But the coal mine might not be as plentiful like it seems here. It was different, anyway, because the way I understood him, they'd work on the level*. I don't know whether they went up or down, but when one level run out they would work another level. Out here you work your one level. Here you have a horizontal seam.

Frank Borgognoni

Borgognoni's story is very typical of Italian miners interviewed here: most worked first in the northern Illinois mines or "out East," then migrated to Christian County at the behest of a relative who helped develop the new Peabody Coal mines.

My parents were raised on a small farm in Italy. My father came to this country to see and locate his father's grave in McAlester, Oklahoma, but he couldn't find the marker. So he came to Ladd, Illinois, right after the Cherry Disaster, and located his uncle. He was sixteen years of age and he worked on his hands and knees, he shot* his own places, he shoveled his own coal, and all that. In order for him to be able to go down below, his uncle had to take him because he was not old enough and didn't have

Divided Kingdom

no miner's papers. So his uncle took him under his jurisdiction. He worked there a couple of years and did not like it. So he traveled to Christian County where he had some friends who were originally from Italy from the same province where he came from. That would be Bologna. One friend in particular, Sam Bernardi, ran the Hawaiian Inn for a period of years in Taylorville, Illinois. Somehow or other—I don't know how—he got a job at No. 8 coal mine.

After three or four years in Christian County, at Mine No. 8, he—through my mother's telling me—saved a thousand dollars and sent it to his mother. She bought eighty-four acres of ground over there. His intentions was really just to work a few years here and go back to Italy. But he met my mother, got married, and he stayed here; he never did go back. When his mother died, they sent my father the papers to sign in order to get his portion of the farm, but he wouldn't take any of it. He said that his brother had taken care of his mother for thirty-five years and that he was going to give his part to his brother, and he gave it all to him.

I had an uncle that lived in Langleyville, by the name of Vincent Corso. He worked at No. 8, and he was responsible for the whole family coming over here, two sisters and a brother. He made the money working in the coal mine and he brought his mother over here. He went down to the bottom one day and got ahold of Mr. Shaw, the mine manager. He said, "I can't work in that place no more. The roof's awful bad, it's going to come in any time." Of course, he talked broken English, and Mr. Shaw said, "You get back in there and go back to work or I'll fire you." He went in there and wasn't in there five minutes before the roof caved in and killed him. It was about a week or two after I was born.

Lena Dougherty

Mrs. Dougherty was born in the northern Illinois mine area to a northern Italian family. She and her husband moved south to the Midland Tract*

I was born October 10, 1903 in South Wilmington, Illinois. I lived there until 1920, so I was not quite seventeen. My father was a miner. He and my older brothers worked in the coal mines up north, which are very low. I have nine [living] brothers and sisters; that's a big family. There was five girls and five boys. I was the sixth.

All my dad and brothers ever talked about was the mines. My husband and his folks had always worked in the mines; that's all we ever heard. It isn't like now where they make the big money. In those days we were poor, but we were satisfied.

My father was a young man when he came from the old country, I would imagine. I don't have any of that data, but I imagine he was in his early twenties. At that time, though, there was a little mine in Clark City. I don't know if anybody has told you this, there were all little mines. My dad worked at first in Clark City and then moved to South Wilmington. Past that, he was always in South Wilmington until we moved to Auburn in 1920. I wish we would have asked more questions about the old country, because my mom and dad didn't both come from the same town. My mother was from San Colombano; my dad was from Busano, but they were both in the Province of Torino. My mother used to talk more about her childhood, but my dad never did. My dad went back to the old country, I would say, about 1930. His mother and dad was still living and he went back for a visit but he never did talk about the old country. My mother used to, but he didn't.

It was mountainous there and they lived, by the way Mom used to tell it, with goats, a cow, and chickens. Her

Divided Kingdom

Figure 18. The newly constructed tipple of Peabody Mine No. 9 in Langleyville, 1917. (Courtesy of the Boch Brothers)

brothers would hire out when they were older. My mother was ten years old and doing laundry, baking bread, knitting and everything. So, the brothers would bring the goats or whatever they had up into the mountains where the greenery was. My mom came here when she was fourteen.

My mom's dad was a stone mason. He contracted pneumonia when she was a little girl and died when she was very young. Her name was Angline, like mine, but Vacca, V-A-C-C-A. My father was Dominick, Dominick Perardi.

When I was a child all the superintendents were all English and Scotch-Irish. I can't remember an Italian being a superintendent. The lower class had high respect for them, men that were above them. I can remember the Fergusons and the different names, they were well thought of. There was no feeling of superiority between those men.

Ada Miller

Ada Miller's Italian family responded to the news of a new mine opening at Langleyville in 1917. The cooperative store that Mrs. Miller's brother managed was established to serve the needs of ethnic miners and became a settlement and clearinghouse of news and job information for newly arrived laborers.

I had an older brother; he was fifteen years old when my father died in 1913. He was born in Italy and was two years old when they came to the States, to America. You know, he was an extraordinary person. He took over like a father. We were living in Riverton at the time during the First World War. He came to Kincaid to work at the cooperative store. It was rumored that they were sinking this mine at Langleyville in 1917-1918, which turned out to be mine No. 9. Mr. Langley owned a lot of land there and at that time it was all timber. Of course, he had to have a grocery store to entice the people to come there. So he started my brother up in business. Bought a building for him, moved it from Sharpsburg. It still stands, if you have been there. My brother was quite successful. That was Mr. Langley's way of getting somebody in there to start the store and this way families would move in. It was just good psychology on his part. He built some homes and the families started to move in.

My father didn't last long after we moved. My father had gone hunting and caught pneumonia. He died at St. John's Hospital in Springfield within ten days. I used to tell my mother later on, when they discovered penicillin, "Well, Ma, if Pappa was alive today we'd save his life." But in those days there was nothing like that, so within ten days he was dead. My mother was six months pregnant with my younger brother, so we really had it rough. I don't remember ever seeing my mother when I would get up in the morning because she had already been out washing clothes in people's homes, in those days. Scrubbing on the board.

Tom Rosko

Tom Rosko's Slovak-American family moved from the western Pennsylvania coal fields to the smaller, non-mechanized mines around Hillsboro and Witt. These mines were developed before the larger, more lavishly-funded mines around Taylorville opened up.

I was born in Dawson, Pennsylvania, in the coal fields near Pittsburgh. My father was eighty-eight years old when he died in 1956. He was born in Czechoslovakia. My dad came here, I guess, along about 1890. He was just like everybody else. He figured [to move here] to Hillsboro because it was a good country to live in. My dad was a cripple and orphaned at about eight years old. A lot of them come and landed in Pennsylvania. That's where the Slovak ones came first.

He came from what they call Slovakia. Our people when they came here were all Greek Catholic and Orthodox. Well, after the First World War there was eighty Slovak families here in Witt. The place my dad came from was all under Russian rule. They either stayed home and prayed themselves or they went to the Roman Catholic Church. Then, I guess, they got enough in their parishes and enough leaders so they built a church of their own. And they had that church going for a good while. They could really chant. I tell you, it would make chills go through you. But you talk about going to church. Now, some of us squawk because we stay in church a half-an-hour. We used to go to Mass for two hours.

I don't know how come my dad landed in Witt, but I guess they had too much work. In Pennsylvania they'd go down and shoot and then they'd load their coal afterwards. There was dead time. So he come to Illinois in 1906.

Leo Waigausky

Leo Waigausky's family first settled in the northern Illinois mining area and migrated to the new mines downstate, in this case in Madison and Macoupin counties. Ethnic families, as can be seen in this excerpt, constituted a shifting, fluid work force.

My full name is Leo Waigausky. I was born July 26, 1910, in Ladd, Illinois. My father was a coal miner. At one time he worked around Ladd, Cherry, Spring Valley, and that area and then he came to Macoupin County. His father came over from the old country, Lithuania.

Around Bureau County—this is around Spring Valley—the coal seam was only about two and a half feet thick. Dad was telling me about crawling around on his hand and

Divided Kingdom

knees, laying down working the coal; but then, that is the only thing he ever told me. Then he came here. I was about nine years old when they moved from ladd. Later on, I went into coal mining.

Before coming to Macoupin County, we moved to Livingston; that is in Madison County next to Macoupin County. We lived there one winter and then came to Carlinville because Standard Oil opened two big mines at Shoburn and at Carlinville. Of course, there were a lot of coal miners who migrated here then. They were both pretty good coal mines, but they only lasted two or three years; then Standard Oil Company shut them down. Then, of course, Superior had four mines here in Gillespie. They had Little Dog* or smaller mines, and mines at Livingston and Staunton. That's what we'd do; when they shut one mine down, we would go try to find a job in another mine.

Figure 19. Typical coal miner's family in company housing near Moweaqua, Christian County, December of 1932. The family is waiting for news on the outcome of an explosion at the Moweaqua mine. (Courtesy *Decatur Herald and Review*)

Family

Otto Klein

Otto Klein was born in Collinsville in 1912 to Germanborn parents who had migrated to the United States as war brewed in Europe. The family worked in the St. Louis-area coal field, which was mined largely by Germans. Some miners found jobs outside the coalfields during lean times and sent money home when they could. When they couldn't, miners' families often relied on generous storekeepers to carry them "on book" until they could pay.

My name is Otto Klein. I was born in Collinsville, Illinois, May 5, 1912. My father came over here first. He was a German; he actually lived across the river in Russian territory where he was an overseer of the land. He designated where certain things would be planted and rotated the crops and stuff. They were allowed their own chickens, their own pigs; the rest of it belonged to the government.

It was a hard life there, but evidently they figured it would be easier to make a living in the United States. Besides that, the war clouds were forming and he didn't want to serve under the Russian rule as a soldier. So he

Divided Kingdom

decided to come over in 1910. He was here about a year-and-a-half, then he had my mother come in 1911. I was the first born of my family here in the United States.

Evidently the area around Collinsville is predominantly German, and someway along the line they must've known that. When he first came here, he came and worked in a leather factory. Then, it was evident that you could make more money working in a coal mine. So he migrated from the leather business to the coal mine.

Now, when my father went to work in the mine, he decided to move from the coal fields in Clifford, Illinois. That's where my father worked, in Clifford. Papa and Mama decided to follow their son-in-law, and Papa moved to Sesser and opened a tavern. He worked in the mine in Sesser and died when I was about seven years old. That left a thirteen-year-old boy, me, and my younger brother and one sister. We lived in Sesser in a company patch. The company owned the houses, and the company store sold you the groceries; it was kind of a difficult task, even if you had a father. It turned out that we owed rent and we was living on food showers for awhile. I don't know where they came from. They came from maybe union people and neighbors or something. I don't know.

I'd wake up in the morning and the table would be piled up with staples, groceries and stuff. There was no one making any money; see, my brother was thirteen years old. He was in the fifth grade. So it was evident that something had to be done, or we was going to end up in the orphanage or something.

And so at that time, it was easy to lie about your age and get away with it. So my brother lied about his age and got a job at the mine at thirteen. He was known as a trapper*. A trapper is a guy who opens doors for motors to come in and out to keep the air coursing where it ought to.

The coal mine would work very seldom in the summertime, so when slack time came, my brother had to go to Chicago and find a job. He got a job in Chicago and couldn't send enough money home, so we were getting behind on the food bill. At this time, the coal company had gone under; they just quit.

A storekeeper uptown, Joe Avati, would carry miners during slack times. So when he went to Chicago, we would still get groceries at Joe Avati's and if he made enough money, he would send it home. Now, the way he got to Chicago was on the train, riding the rails as they called it. So when the mine started getting back to work like it should, he came back and took care of us. My brother got a job at Buckner, at the Old Ben Mine Two. He worked there for awhile and then when I got out of school at the eighth grade, I said, "It's time for me to take the family and support and you can get married." He said, "No, I'll send you to high school if you want to go." I said, "No, it's my turn." He said, "Well, we'll both stick around." So then I got a job at Buckner.

Alvin Amuel Wise

Wise was born in Taylor Springs to Irish and German parents in 1913. Alvin and his brother worked in the mines around Taylorville all of their lives. The family was very poor in material goods, but there existed a family closeness that bound them tightly together.

I grew up in this home full of kids; my mother had ten. I am the oldest. I had to go to work to help. I am not trying to put any glory on myself now, but I had to help raise the rest of the family.

I never had much. I don't want you to think harshly about my mother and dad; they were good people, real good. But what I am going to say is, well, they weren't very

Divided Kingdom

fair to you. I used to work two weeks for a dollar of pay; that is all I took. The money went to the family. Not to the parents, but to the family—the operation of the family. I gave it to my mom and dad. Mostly to my mom, being that my dad wasn't an educated man. She paid the bills and done the grocery shopping and everything like that. That is why I felt sorry at times for my mother, because on pay day when a bill would come due, she was the one that had to go and say, "Well, we haven't got it this pay." You understand what I mean? And I always felt sorry for Mom.

All my brothers and sisters was two years apart, two years and six months. My brother, Tony, passed away, he was born on the 15th of October, two years after I was on the 6th of October. It went down the line like that, two to two and a half years. I am the oldest, I will be seventy-three years old on my birthday [1986].

We lived in a little shanty. There wasn't nothing around but shanties in Tovey. The company houses were in Kincaid. There was no company houses in this town of Tovey. These were built by the miners. We had about three bedrooms for seven boys and three girls. The boys was pretty close to staying in one room.

My dad and mom never had no money. When payday came, whatever they had had to be paid out—to pay their bills. We put our money into the house. It is not a mansion, but still serves our purpose. We don't need nothing else; this is enough for us.

When we lived in Jeiseyville my dad walked to No. 8 to work, back and forth. I seen nights he came home, between No. 8 and Jeiseyville. It would rain, and he'd come home and every stitch of clothes on him would be froze stiff. I never appreciated them things but I can sit here and visualize the hardships that my dad did go through for me and my brothers and sisters, and my mother. I have to appreciate him and love him for doing that. My dad never let his family go hungry ever. No matter what little we had, there would be something to eat.

My mother showed affection in ways. I know she loved us and my dad too. I know my dad loved us because he suffered a lot of hardships for us. But as far as an outright showing of affection, my dad never did do that, ever.

I remember one time my brother got burnt badly when he was seven years old. We lived in Pawnee at that time. His leg was nothing but raw flesh. She lay him in the sun, and honestly believed she healed him herself by the sun. She lay him there with no clothes on and she had a bed affair fixed there. I remember when I was a young boy I used to have terrible leg aches. My mother would stay up sometimes most of the night rubbing my legs. That is affection and love; it ain't just because you have to put your arms around somebody. But my mother showed affection, yes.

My mother, until she and my dad retired, had a hard life. Everybody had hard lives when I was young. Nobody had an easy go because there was no work. No jobs, half enough to eat. My mom and dad showed their affection in ways.

My mother when she would get mad would grit her teeth at us. I have never seen my mother mad at me, ever. But my dad would knock you down and kick you for falling if you didn't act right. Of all our brothers and sisters, I haven't seen no backbones come out of us.

A typical Christmas when I was young? We had stockings hung on the wall. We would get an orange if my dad was financially able. We would have an orange and an apple and a few Christmas nuts and a little bit of hard Christmas candy and that was it. No toys—maybe once in awhile if he was financially able, we would have some toys. In later years, when my younger brothers was say ten, eleven, twelve years old, we might get a pair of overalls, but that wasn't very often. I only had one suit in my life while I was at home. Come to think of it, I didn't like it either.

Divided Kingdom

I remember the time my dad and my uncle—one of my mother's brothers—saved all summer because we knew the mines was going down. They would save up; they would buy flour and lard and things that would keep. We would go down to the South Fork River and stay there practically all summer. Just live down on the river in tents. Eat fish, squirrels. If you didn't save any in the summer, you didn't have nothing in the winter. That was just sort of a habit. Way back there in them days we knew it was cheaper and everything than living in a house.

Lena Dougherty

The youngest children in mining families worked hard. The older boys worked in the mines and surrendered their money to the family for its survival. Poverty was shared by all mining families: they were only acutely aware of their misfortune, however, when the world outside remarked upon the extent of their poverty.

We did anything for work. For chores, of course, the dishes, the kitchen floor was swept three times a day—after breakfast, after lunch and, of course after what we called dinner-supper. (Laughter) We helped our mother wash clothes. Of course, you washed on the board; we didn't have a washing machine. We took care of the kids that were younger than us. My oldest sister was ten years old when I was born and she could wash clothes or cook a meal at that time.

We didn't have clothes like they have now. You had two dresses for a week day and then would have one for Sunday (laughter). You didn't have any clothes closet. This house had no clothes closets. Now what would they do without clothes closets (laughter)? But we had two dresses for everyday. You washed one and you wore the other.

We used to just love those old ladies. I never knew my grandma because my mother came over to this country when she was fourteen. But they used to amaze me when we would go over to visit people. The babushka* would be sitting behind the stove with her rosary. See, we had more Polish people in our neighborhood than Italian. There were quite a few Italians over on the north side. Of course, we were all neighbors, but on our side there was mostly I would say, Polish.

My mother never talked Italian to us. She came to this country when she was fourteen years old and got married when she was eighteen. My dad and mother talked Italian to each other. My older brothers and sisters could talk it and understand it. Just so you can understand what's being said that's the main thing.

Like I told you, all the kids in those days gave their money to their folks. My husband was the same way then. He worked in the mine and all his money went to his mother. She gave him the last pay that he had; he could have it before we married.

So that's what we got married on. Part of that he threw in when they shivareed us. So we had nothing. My mother, of course, had a hope chest, you know, with the sheets and the pillowcases and everything. But that was the sum total. So we borrowed two hundred dollars from my dad and bought our furniture. After you paid that off, if you could afford something else, you got more. No, we didn't have anything straight away. (Laughs) That was in Auburn where we lived first. We lived in a little three-room shack. We only had two rooms furnished. We had the kitchen and the bedroom furnished was all. We didn't get any more furniture than that. After Jean was born we lived in Tovey and we didn't start buying furniture until we had Ruthy. That would have been after we were married about five years. Then I think we got

Divided Kingdom

a living room set; but, no we lived poor. But everybody was in the same category so you didn't feel like you were missing out on anything, you really didn't. We had good times.

We moved to Tovey because my husband got work there. When he first worked at Tovey we lived in two rooms. But he got his job in January and the miners would be out on strike in April. They always went out on strike in April, so we would only be there the three months. So there was a family in Tovey that rented us two rooms for our kitchen and our bedroom. Then when April came, we went back to Auburn. They were off working from April to October. He would ride back and forth to work in Tovey.

That first winter there was so bad (1936); the roads were so bad he couldn't make it with the car. He had to board for a month or two in Tovey and then we moved back to Tovey again. We stayed in Tovey until we moved to Kincaid during the mine trouble. We have been in this house for fifty years. It will be fifty years in November. It was 1936 we moved here to this house.

Frank Boch

The Boch narrative illustrates how hard everyday life was for coal-mining families during the Depression. While middle-class and white-collar workers most typically did not cultivate a garden nor butcher their own meat, miners' families had no choice. The Boch family, for example, survived on their garden and livestock and produced the majority of foodstuffs by canning, baking, and butchering.

I remember my mother cleaning house and we didn't have carpets like you got now in these modern houses; it was just the bare floor. They might have thrown a few rugs down or something. Every week she'd get the bucket

Divided Kingdom

Figure 20. The Boch family, like other coal mining families, survived on their gardens. Stephen Boch took this photograph of his son Frank in the strawberry patch behind their house in Hewittville, in 1911. (Courtesy of the Boch Brothers)

Figure 21. Boch brothers, Fritz, 2 1/2, and Max 4, sit beside baskets of strawberries picked from the patch, June 1, 1914. (Courtesy of the Boch Brothers)

with soap and scrub the floor with the scrubber. I can smell that soap and everything yet. She kept it clean.

We didn't have much clothes. When we went to school we put on a pair of clean overalls. Most of the kids wore overalls, and when you come home from school you put on another pair of overalls that was older or patched up to play in. Then you put the same ones back on the next day. We didn't have clothes like we do today.

We managed washing and bathing the old-fashioned way; you took a bath in a galvanized tub. In the summertime, I remember, in the morning we would pump the tub full of water and let it sit in the sun and in the evening you would jump in and take a bath. In the wintertime, of course, you would have to heat the water on the stove. But my mother done all the washing, I remember, on the scrub board.

We had a lot of food that came from the recipes from Austria. It was simple food, that is one thing I can say. As many hard times as we went through in our lifetime, we were never hungry. We had food to eat even if it was simple. I remember my mother would make some kind of cornmeal mush. Get the cornmeal and mix potatoes with it and put some bacon on top of it; that was a meal, it filled you up. You could do a day's work when you ate that; it wasn't bad. We always had milk from the cow, and chickens, and hogs. The folks raised hogs and butchered them. We had our own meat and they would make their own sausage and everything. They would grind it up by hand. We butchered in the wintertime. We didn't starve. But there was other poorer people than us and I will tell you they starved.

My mother always baked bread and she made jelly and canned. They done a lot of their own work. We've got a good picture of mom with a loaf of bread that she made.

We never had much money, but my dad didn't drink. See, a lot of them coal miners drank their money away on pay day and frittered their money away. They were just as poor as could be; they didn't have anything. They throwed their money away. My dad didn't do that, he didn't care about the drinking. He bought a camera, took pictures, and had a lot of guns. He watched what he was doing. If the mine went down most miners would be broke next week.

We were brought up a certain way that you bought what you could afford, that is all. When I first got married I remember my wife started wanting to buy stuff on credit and I was against it. I didn't believe in credit because I was brought up different, you know. I thought for sure we were going to be bankrupt or in the poorhouse or something, you know, when we was on credit. I didn't think we were going to make it.

Figure 22. The Klondike Hotel in Taylorville was typical of the restaurant/bars in Christian County during Prohibition. This scene was preserved by Stephen Boch's camera in 1925. (Courtesy of the Boch Brothers)

Divided Kingdom

My parents would send us kids to the store but then it was a different time. I remember they'd give you a nickel—a nickel was hard to get—and you could go to the butcher shop and you get a great big soup bone. It was as big as a football or bigger and a lot of meat on it, they didn't trim it like they do today. My mother would put that in a pot, put some noodles with it and potatoes or whatever and we had a good meal out of it.

We could talk to my mother because we were close to her, but my mother never did learn the American language very good—she could talk some, but very little. My dad could talk pretty good. And he could write the American language. He was a good writer, but my mother never did catch on. Relatives and close people could talk to her and she'd translate.

Frank Borgognoni

Depression and poverty demanded discipline to enable the family to survive. Frank's father dished out strong discipline and demanded that everyone pitch in with the household chores. Housework and child rearing wore women out; Mrs. Borgognoni had two nervous breakdowns. In mining families the women managed the money. They had a better idea of the expenses, and didn't drink as the men did. In the coal towns the entire neighborhood disciplined the children, and helped out when work needed to be done. The neighbors, in these instances, acted as an extended family, enforcing family rules and reporting violations to the parents.

I would say that really we were not close to my parents. I respected them both and I had to treat them both alike. I couldn't say, "Hey, Mom, Pa got mad at me and he's going to do this." You know what she'd say? "I'll give you some, too"." Because they stuck together. You didn't dare tell my dad that you was really mad at my mom and she did this and that, because he'd bop you one. He wouldn't fool with you. My dad hit me once or twice in his lifetime and I never forgot it. For instance, one time around Halloween, he called me to the end of the bar and said, "I want to talk to you." He had a hand on me that was that big. He pounded it into that bar and said, "Don't let the law come down there looking for you. If they get you and put you in jail, I'm coming after you. I want you." He didn't have to say no more. He said, "You know I don't chew my tobacco twice." I said, "I know you don't." That was it. I was fifteen years old.

I was talking about how my dad kept one of the most well-kept joints in Christian County. They'd get a little loud and he'd just go like that on the bar [knock on the bar] and everything would quiet down. There was no more noise. He was just that way, he just wouldn't tolerate no monkey business. A woman came and told my mother, "My husband can come into your tavern any time he wants to because I know there's no trouble and no women in there." See, at one time, there was never no women in my dad's tavern, never. It was just a man's tavern. That's all it was. And no profanity in there, nothing. It was unbelievable.

My brother cussed him one time when he was thirteen years old in front of the tavern, and my father didn't say nothing. It made me kind of sore. I thought, "Well, if that had been me, he'd have really laid it on me. Here my brother does it and he gets by with it." So the day progressed on and that evening, about five o'clock, he came to me and said, "I want you to take care of the bar. I've got something I've got to do in the back." I heard him close the door—we lived in the back then—and he locked the door; I heard it click. I never heard so much commotion in all my life. He had my brother back there, and I mean he worked him over good. My brother never cussed him out no more; that was the end of it.

Now, my sister was no trouble at all. My sister, till this day, don't smoke, and she don't drink. You'd have to meet her; she works up there at Frank Switzer's. She's one of the type of people on the up and up; unbelievable.

Well, with my parents, you just didn't get out of line, and you didn't say, "No, I'm not going to do it." If they told you to get something, and you couldn't get it right then, you'd say, "Would it be all right if I get it in a few minutes?" It was either yes or no and that was it.

Even as a boy, I helped my mother in the house. I've done dishes. Me and my sister used to help her clean house—all that sort of stuff. We knew she was tired, my mother was sick a lot, see. She had five major operations in her lifetime and two nervous breakdowns. She even had a miscarriage. If you had to live like we did then, you'd understand.

You didn't have hot running water in the house. You had a copper boiler which I've seen my mother lift off one of those old Majestic stoves about two-thirds full of water by herself—just lift it right up on the stove. That was hard work for a woman. At nighttime, we didn't have electricity when we was in Tovey. We had coal oil light, or kerosene light. She'd sew socks and stuff by kerosene light. You didn't dare let your mother carry a bucket of coal in the house. You just didn't do it.

We felt close to Mrs. Sassateli who lived in Hewittville, She was like a second mother. One of the nicest women in the world. I'll never forget when she died how my sister cried. She said, "You know, Frank, that's our second mother." That's just how close she was. Of course, she was related to my dad; she was a cousin of my dad in Italy.

I had one grandmother in this country and that was it. I was close to her. I was the only kid that was close to her. When the weather in the wintertime got real bad, that's

Divided Kingdom

when I was six years old, we moved to Langley so my dad could get to Mine No. 58 in Hewittville a lot easier. See, in them days, we didn't have [Highway] 104. You had to go through the country.

Even at holidays my dad spent his time uptown and my mother at home. That's just the way it was. He usually went uptown; that's when they had them places on Hump Street in Tovey. They were more or less bootlegging in the open in them days. They had Viking Tavern right there on Front Street. They had a poolroom where you could always get a beer, called it a mule. It was bootleg whiskey.

My mother went to church. My father never did go back to church in this country. I don't know why. We didn't go. My sister went for awhile and she got married in the church. When my grandmother found out that my mother got married and didn't get married in church, there was a lot of hell. She came right down and got ahold of me, my grandmother did, and baptized me. That's just how she felt about it. I went to catechism for awhile. I wanted to be like the rest of them.

There was never any arguments about money. When my mother needed money, and my dad had that tavern, she went to the cash register and took what she needed; he never questioned her. I never seen my father carry five dollars in his pocket. That's unbelievable.

You see, you had a mixture of foreign class of people and they didn't all operate the same in their families. Comparisons are difficult. There were good and there were bad. But all our people wanted us kids to respect everybody and that was the end of it. They didn't tell you the second time. In other words, flat, straight up. God only knows what would've happened if some follow came in and said, "Hey, Pete, your son did so and so." "You sure?" He'd come and ask you if you did it. "Tell me that you did this." Just like one time when I was a kid in Tovey and I bought a broken down scooter that you scooted with your foot. My mother said, "Where'd you get that?" "So and so gave it to me." Okay, there was no more said about it. Pretty soon, about fifteen minutes later, I seen my mother walking down the sidewalk. Of course, she went down to so and so's house and asked him, "Did you give my kid that scooter?" They told her, "Yes." Just as long as you don't steal it. You see?

In the small towns—Tovey, Bulpitt, Kincaid—we all mingled together. Even the older people. I noticed that we had a lot of French people in Tovey and they got along so good. For instance, I want to recite something here. Just like when my dad had to dig that basement, he didn't say, "Hey you, John" or "You, Edward, how about you coming out and helping me." They found out my dad was going to build a basement and here they come with their own spade and their own wheelbarrow, fourteen of them. But you better provide them with plenty of wine and something to eat, or else. Really.

I watched my mother when she started to get that dinner ready. Here come the neighbor women. They got to helping her, too. And hey, when the neighbors got to digging something, you seen the old man loading his spade and his wheelbarrow. The old lady didn't say, "Hey, where are you going?" There was no question about it. I'll never forget when they dug that well.

In Tovey it was mostly Italian. They got along with them Lithuanians real good. They worked together down in these mines. This guy was your buddy; that guy was your buddy. The miners used my dad for an interpreter. Even these superintendents when they had somebody they had to tell or the mine manager had something to tell the guys down in the mine—if there was a bunch of Italians—my dad was there. He'd convey it the best way he could; my father was good at it.

Community

Jack Batuello

Miners had bad experiences with all representatives of authority—the tax collector, sheriff, judges, and the coal company. All they had to rely on was the solidarity among themselves.

I can explain our conditions better by saying that we were not even free citizens in our community. The company in the early period owned the houses in the community [Kincaid] and they owned the company store. In some cases they issued their own script for money. The prices at the company store was about fifteen, twenty, and twenty-five percent higher than in a private store. Even the priest, the tax collector, the sheriff, and the judge, all belonged in the pocket of the coal company. We were virtual prisoners. And so, when we lamented too strongly about the bad conditions, we were demoted to a worse job.

Divided Kingdom

Figure 23. National Guard troops, called in at least four times over a four-year span, were part of the power structure. Here they prevent the PMA strikers from meeting in Manners Park, Taylorville, September 27, 1932. (Courtesy Decatur Herald and Review)

Joe Ozanic

Joe Ozanic's family lived in company housing in Pawnee. All immigrant mining families were forced to live in the "Patch" and were looked down upon by the rest of the community. Liquor defined the major differences between native born Americans and the various ethnic groups living in a community; miners simply drank, and the natives didn't.

The Patch in Pawnee—that's where all the foreigners lived that immigrated over here. They might have a cousin or an uncle or some relative here and they would be invited to come to work in the mine and the majority of them all lived in the Patch. It was just outside the city limits, due to the fact that they was backward people. Some [native-born] people thought maybe they was

Divided Kingdom

Figure 24. A baseball game played in Pawnee's "Patch," 1925. Notice the company homes in the background. (Courtesy of the Boch Brothers)

Figure 25. The Star Trading company, pictured here in 1904, was Peabody Coal Co.'s mining store in Pawnee. Almost all the miners' clothing and supplies were bought at the company store on credit. (Courtesy of the Boch Brothers)

just a little bit better than these people that was coming over from Italy and Ireland and England and Scotland—and the Polish and Lithuanian were all mixed here. They kept that end of town out of the city limits so they couldn't vote to control the village. That was done for just one purpose: they wanted a local option which controlled the booze. They didn't want them to vote because they knew they would vote the town wet. They were used to wine and beer and liquor and the natives wanted it dry.

They had a company store here and they put out script. I could sign up and get a two-dollar-and-fifty-cent book of stamps, a five-dollar book, or a ten-dollar book, and I would go in there and buy groceries and give them food stamps. They would call up the mine and ask them, "Well, so and so wants a ten-dollar sticker. Has he got enough money coming in to cover then ten-dollar sticker?" The Star Trading Company's prices were not high—they competed with the older grocery store.

John Ralph Sexson

Those miners who lived in Kincaid were fortunate. House rentals were affordable and the houses themselves were spacious in comparison to other housing available to miners. As inexpensive as the housing was in Kincaid, however, the miners still could not afford the purchase price. Selling bootleg was a way to make extra money to afford house payments.

Peabody built a lot of these houses to start with at Kincaid, company houses and all. The biggest part at that time were company houses. I didn't live in one at first when we moved to Kincaid, but I did later on. In fact, the one we live in now was a company house. We done a lot of building and remodeling and stuff like that and they're all nice homes around Kincaid. At that time Peabody started building those houses in the early 1920s, I believe. You could buy a five-room house, if I remember right. Fifty-eight dollars—a very, very low price, if you could do it. We didn't have the money at that time; you couldn't borrow like you can now. We couldn't even borrow enough money to get a down payment on a house until 1941 or 1942. You just couldn't go out and borrow money like you can now.

There's a lot of drinking going on in the coal towns. I don't think it had anything to do with the work conditions or anything like that. Italians at that time were wine drinkers, always have been; and Americans were more whiskey drinkers. Beer or whiskey drinkers. During Prohibition in Kincaid, Jeiseyville, Tovey, and Langley almost every other house was a bootlegging joint. I used to go there when I was a teenager—to Kincaid—to these bootlegging joints where the liquor was sold, and that's what they done for extra money.

Figure 26. Young rock pickers from Peabody Mine No. 58 at Kimlake in Christian County, 1926. Harry Filson was the picking boss. (Courtesy of the Boch Brothers)

Divided Kingdom

Louis Wattelet

The mining community pulled together in hard times in all Christian County towns. The women fed the unfortunate, nursed the sick, and worked as a team to repair, paint, and even construct homes.

Well, we had a little four-room house in Stonington. Our light bill was a dollar a month. My wife's mother was still on the farm and she would give us chickens and stuff like that. We all managed; nobody had any money. People were close in them days, you would be surprised.

You know they say "the good old times." I think what they mean was that people were concerned about one another. You know, everybody was in a hard way and if somebody got sick the whole neighborhood was over to help out. Now, you don't even know your next door neighbor anymore. It is strange, isn't it? During the influenza following World War I, my mom and other women in Stonington worked with the doctors from home to home. They would make a pot of soup for that family, made their rounds every day, made sure they ate and everything. Mom never did get the flu and us kids either. Isn't that strange? She worked into it and we didn't get the flu; that was strange, wasn't it?

If the landlord would get the paint, we'd paint, and keep the houses up. We were all good people. Nowadays, say, let the landlord do it. I would say, "Landlord, get me some paint so I can paint this place." I wasn't working anyway, so what the heck.

Figure 27. Employees of the "glove factory" on East Main, Taylorville, mid-1920s. Most of the women were wives of Peabody miners. (Courtesy of the Boch Brothers)

Figure 28. One outlet for the miners was playing in a locally sponsored band. Nearly all members of the Stonington Concert Band in the mid-1920s were coal miners.

Divided Kingdom

Frank Borgognoni

The "Patch" towns were a heterogeneous mixture of all nationalities and creeds. Nevertheless, the community pulled together to accomplish most of the work, as Frank Borgognoni relates.

Now, I was one of the fortunate ones. My father had saved enough money—he was working just that short time in the coal mines—that he had Mr. Yearly, a carpenter, build him a house. The Italian fellows helped him build the basement, they dug it by hand. When Mr. Yearly and Mr. Westbrook got done with laying the blocks, the concrete and building the house, my father owed Mr. Yearly $1,900. He told him, "Pete, your house is done. It's all yours." He said, "How much do I owe you"? He told him, "Nineteen hundred dollars." He said, "Don't go away, I'll be right here in about two minutes." Came back, gave him $1,900 cash and paid him. We had a new house when we lived there. The house was built in 1928, I believe. I was about four or five years old. I was just young enough I could just barely remember it. The Italians helped to build it.

Ninety percent of them was Italian in Tovey and in Kincaid. Bulpitt consisted of ninety percent Lithuanians. Kincaid had just a duke's mixture, just everything. Johnny Bulls, English, that's what the Johnny Bulls are, Frenchman, Italians, just all kinds of people in Kincaid. Irish; we had a lot of Irish people in Kincaid. I always kind of thought that the English was about the biggest percentage of the nationality in Kincaid in 1920, 1921. That's when all those coal mines got started. They all lived in Peabody company houses they had built across the tracks in Jeiseyville. Italians came mostly from Rome on up. From there on down, it was very, very rare. That's

Divided Kingdom

Figure 29. The Merchants' Bank, 1923, sponsored by Pawnee merchants. (Courtesy of the Boch Brothers)

Figure 30. An unusual photograph of the Miners' Band underground in the Superior Coal Company's Wilsonville mine, 1937. This band was funded by Local No. 1, PMA (Photograph by Pathe Services)

what Langley was full of, and they had a few Frenchman there. The Italians understood each other's language and they got along fairly decent. But I talked to a fellow one time. I said, "Why is it that you people came to these foreign parts, to Peabody coal mine?" He said they needed a job bad and they were good workers. That's the story that I got—excellent workers. They didn't come down there just to goof around, and they worked every day. You didn't have to worry about them not working. When that whistle blew they'd be there. It was something they didn't have to worry about, the absenteeism, which is a great factor in any of your industries today.

What pulled them together were activities like even on a Saturday or Sunday when they weren't working, they were off, they would always get together. Say, well, let's go get so and so, and they'd get out and play bocce ball or they'd roll that big long cheese. They got along, they'd get drunk and get in a few arguments, which is only natural, it's going to happen. It happens in every race of people.

Tom Rosko

Initially, between 1909-1919, work in the Midland Tract and in Montgomery County was plentiful. Everyone had stories about established households taking in the surplus mining population as boarders. Beginning in the late 1920s, however, communal and family networks got men jobs outside of the coal towns in other industrial centers, and got out-of-work miners "on book."

You wouldn't believe this. At one time, and records will show this, there was over five thousand people in this town of Witt between 1906 and 1915. And then the bottom fell out. There used to be two mines here. But then Nokomis was working and Coalton was working, between 1909 and 1915. It was so crowded people were doubling up. I can remember, there was a cousin of my wife's who come from Pennsylvania and couldn't get no work in the mines there. And he got himself a job here in the coal mines. We doubled up and I think there was six of us in the family. All they had was just a sleeping room and they cooked. There was a lot of doubling up. But that only lasted about six years. Everybody moved out when the mines closed. They all went to the cities. See, that's when the cities were opening up. A lot of factories with the steel work, steel mills, and things like that.

I can remember when we used to go out on strike in the summertime; under the United Mine Workers we'd go out on a strike for maybe two or three months. The mines under the UMW used to shut down until about 1930, 1931. We just lay around. Do nothing! Except them days around here the biggest part of the people raised their own cows and hogs. They butchered in the winter. I can remember during the height of the Depression when we used to work only about one or two days a week. I never will forget. We'd go out here in the wintertime, have to cut wood and everything else. They'd get together and bring their own lunch—what have you—and bring black coffee.

But when the coal depression came the biggest part of them when they got sixteen years old didn't go to the coal mines. They all went to Cicero, to Chicago, Pennsylvania, or stuff like that. I'm the only one that stayed around here. They went there because there was no work around here. They got jobs. You take them days there wasn't much money. My brother said they used to make maybe fifteen, twenty dollars a week way back in 1916, 1917, and 1918 working in them factories in Cicero.

You see, in those days you only worked when they got

Divided Kingdom

the orders. Very seldom did we ever work five days a week in the coal mine.

My wife came from a coal mining family. What I mean, her folks didn't have anything either. We got married in a rough time because I was only working two or three days a week. As I say, the good Lord was good with us. We lived fifty-five years together. She had a rough time. Here's what I mean by that. She worked at the glass factory in Hillsboro. Well, they were always wanting help. We raised the three boys but she never did go to work until she put them in their first year of school. And them days during the war they had a factory way up here about eighty miles in Illiopolis where they made bullets and stuff like that. She done her part of the work. Five of them drove all the way down there. They drove the car every day. She wore our car out. She used our car to do all the driving.

In those days, I'm not ashamed to tell you, there was no money to handle. You're working two days a week. And, you know what, if it wouldn't have been for the credit in them days I think we'd all starve to death. All the storekeepers kept us "on book." In them days, you know, they had a butcher shop. So this was 1936. We was only working one or two days a week and that was on division of work*, see. So I went to the butcher shop. And this fellow, he said, "Tom, your bill is getting pretty high." I said, "Yes, I realize that. And I don't know how I'm going to pay you." Three months after that Coalton opened up and after that I was honest enough to pay my bill.

You take these stores around here; they all carried us. My wife was in the hospital with a fellow by the name of Mack, a storekeeper. He carried all of them in Springfield, John Mack!

Divided Kingdom

Before we went "on book" they had miner's cooperatives. You see the Slovak way back in 1913 to 1914 had a cooperative store and they went busted. They needed management. Too much credit, I guess. Then a fellow by the name of Butchko; he bought them out. That was way back in 1919.

Divided Kingdom

Figure 31. Underground timbermen in Peabody Mine No. 58, mid-1920s. Notice the cloth caps, carbide lamps, the timbering job done by the two men shown here, and the debris on the floor. (Courtesy of the Boch Brothers)

CHAPTER II
Work, Safety, and the United Mine Workers: The World of the Underground Miners

> *That used to be a joke in those days (when we coupled coal cars)...that a finger's a new Chevy, when Chevys were about $500.00. So if you lost your finger, you got compensated and you could buy yourself a Chevy.*
> —Stuart Lidster, 1985

Early Mining and Mechanization

The work in the Peabody mines before mechanization in the 1930s—the drilling, blasting, timbering, and loading—was dangerous, tedious, and exhausting. The key to survival was the miner's skill in negotiating a series of complex, labor-intensive jobs without accident: undermining the seam of coal*, blasting the pocket*, and loading the coal into cars at the face*. Making a living and the ease of the work depended upon a number of factors beyond the miner's control: the existence of good "top" or ceiling, thickness of the seam, and especially the availability

Divided Kingdom

of cars in which to load coal.

The first job of the miners was to undercut the coal at the face. With a pick the miner made a horizontal slit a few inches above the floor and cut about six feet into the seam. Before machinery was introduced this task took two-and-a-half to three hours.

The next step was to drill a hole in the coal face. At first, miners used a five-to-five-and-a-half-foot long auger drill, turned by a U-crank. By the 1930s at Peabody's newer mines the miners used air-powered drills to accomplish this task. The miners would drill a hole into the face a few feet above the cut and fill the hole with powder. The subsequent blast would break the coal loose below the hole.

The third step was to shoot down the coal with black powder or explosives. The miner put the powder in the hole through a needle in the drill. He then packed in his "squib"—a thin roll of waxed paper with a little powder in it—in the entrance of the hole, then lit the end and ran to safety. The cartridge or squib exploded and a ton of coal was shot down and away from the face. Shooting the coal down was especially dangerous and demanded skill in placing the shot. If the shot was improperly placed or fired, the roof would fall on the miner.

The last task was to load the blasted coal into cars at the face. While the miner and his helper shoveled the coal into the cars, they had to remove pieces of rock and slate to avoid having their pay docked for dirty coal. When finished, the miner pushed the loaded cars out of the room to be hauled to the shaft by a mule or trip rider.* In the early part of the twentieth century, the miners had to lay track from the entry to the face and timber the room* before beginning work. In the 1920s, special company men laid track and timbered.

The miner was an independent craftsman before the coming of the loading machines. He worked without direct supervision while possessing a monopoly on a body of

Figure 32. Two miners with their auger drill in Mine No. 58 in the mid-1920s. The drill bit rests on the face. (Courtesy of the Boch Brothers)

Figure 33. Miner (on far right) and his two helpers shovel coal at the face, Peabody Mine No. 58, mid-1920s. (Courtesy of Boch Brothers)

Divided Kingdom

Figure 34. Two miner's helpers stand by a buggy loaded with coal at Peabody Mine No. 9, Langleyville, April 19, 1916. (Courtesy of Boch Brothers)

Figure 35. Miner's helpers and buggies loaded with "flat" coal, Peabody Mine No. 58, mid-1920s. (Courtesy of Boch Brothers)

Divided Kingdom

Figure 36. Fully-loaded cars or buggies waiting to be hauled by mule from the face to the shaft. Moweaqua Coal Company Mine, 1933. (Courtesy *Decatur Herald and Review*)

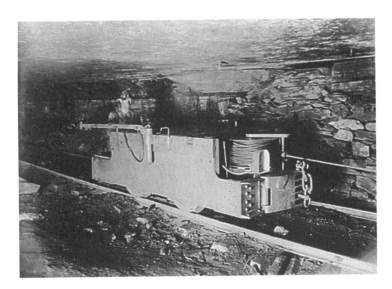

Figure 37. A trip rider underground at Peabody Mine No. 9, Langleyvile, early 1930. (Courtesy of Boch Brothers)

Divided Kingdom

Figure 38. A Morgan-Gardner cutting machine taking a "sampling cut" on a "short wall," mid-1920s. Notice the bars holding up the ceiling.

Figure 39. Morgan-Gardner cutting machine moving into position at the face, mid-1920s. (Both photos courtesy of anonymous Peabody miner. Donated to Illinois State Historical Library, 1988)

Divided Kingdom

knowledge which he passed on only to fellow miners, and kept his own hours. At the same time, he was exposed to risk and accident, and management considered him solely responsible for his own safety.

From the very beginning Peabody Coal was an early pioneer in mechanizing its mines, deskilling* the miners craft and redistributing the mining work to other underground workers. The first machines the company introduced were mechanized substitutes for the manual task of undercutting the face. These cutting machines moved on a four-foot-long cutting bar or chain that extended from the machine forward into the coal face at ground level. At the end of a cut of five to six feet in depth, the miner withdrew the machine, moved it laterally to make the next cut, and repeated the process along the length of the coal face. By 1925 over ninety percent of Illinois' mines, including all of Peabody's operations, had cutting machines.

Though the machines were welcomed and time-saving, they created a new danger. The machine produced coal dust that led to a combustible situation underground, and also increased the risk of miner's asthma or "black lung." The costs of improvement were borne by the miners through additional safety and health risks.

After the cutting machine was introduced, mining in the Midland Tract was "deskilled." Many of the underground tasks formerly handled by miners were handled by labor gangs. In the 1920s, Peabody hired "company men" who specialized as shot-firers, track-layers, drillers for shots, and timberers. All were taught their skills by company engineers and managers. Instead of piece work, these men earned their money on a daily wage directly from the company.

Deskilling was a significant step toward mechanization, since it allowed some of the new men with no mining experience—mostly immigrants—to do the gang and mechanized work. Consequently, the composition of the total work force

Divided Kingdom

changed. Miners/loaders declined from ninety percent of the underground workforce in the early 1920s to seventy-five percent or less in Illinois by the end of the decade. By 1930, the composition of the Peabody workforce in Illinois was:

```
Machine Miners . . . . . . . . . . . . . . . . . . . 8,123
Machine Runners . . . . . . . . . . . . . . . . . . 779
Trip Riders . . . . . . . . . . . . . . . . . . . . . . . 373
Trip Rider Assistant . . . . . . . . . . . . . . . . 339
Shot Firers . . . . . . . . . . . . . . . . . . . . . . . 203
Foremen, Fire Bosses . . . . . . . . . . . . . . . 123
Trackmen . . . . . . . . . . . . . . . . . . . . . . . . 455
Timbermen . . . . . . . . . . . . . . . . . . . . . . . 184
Pumpmen . . . . . . . . . . . . . . . . . . . . . . . . 160
Total . . . . . . . . . . . . . . . . . . . . . . . . . . 13,681 *
```

Source: Keith Dix, Work Relations in the Coal Industry, p.11

Figure 40. A Morgan-Gardner loading machine, called the "hog" by the miners, is wheeled into place in one of the Peabody mines in the 1930s. (Courtesy of anonymous Peabody miner. Donated to Illinois Historical Library, 1988)

Figure 41. Mule-driven coal, Peabody Mine No. 9, Langleyville, 1917. (Photo courtesy of the Boch Brothers)

Figure 42. Underground miners and employees of the Christian County Coal Company, circa 1905. The men in the center and left are mule drivers who hauled the coal out of the mine. (Photo courtesy of Brenda Henning)

Divided Kingdom

The world of the underground miner in Illinois was drastically altered beginning in the late 1920s when the larger coal companies—including Peabody Coal—began introducing time-saving and man-hour eliminating machinery to load coal. The new machine, a mechanized coal loader, was initially developed by Peabody engineers. Most important, the work force was halved at sites where these loading machines were introduced. Only a few men were needed underground to drill, fire shots, and move the heavy loading machines. In Christian County younger, more agile men replaced the older miners in tending the machines.

Harry McDonald

Harry McDonald entered the mines in the early 1930s as a hoisting engineer. Here, he relates what immediately preceded mechanization of the mines in the 1920s.

At that time, in the early 1920s, most of the transportation was what they called mule driven loads, you know; they brought their coal out with these mules. At that time, they had in some of these mines up to thirty or forty mules that they kept underground all the time. When they shut down the mine in the summer they would bring them mules up; and they'd have to cover their eyes when they brought them out so that the sun wouldn't bother them. They would go blind if they didn't. They'd bring them out because they'd just get fat and would create a work problem underground. They didn't have the motors* that they had in later years; this was before the electric motors came down in the mine. They didn't have electricity down there either in Mine 58.

This was all hand-loading, and these people had to work with a carbide lamp. It was always hazy, you know. They

Figure 43. Road motor in Peabody Mine No. 58, 1920s. (Courtesy Peabody Coal Company)

were looking through a haze all the time. They had to drill holes themselves in the coal face to load with powder and it was always a dangerous project for them. Men would drill these holes and load them with black powder, and towards the end of the shift, there was a fellow that came around—they called him the shot firer—and he would inspect to see if the hole was okay, tamped and everything so that it would shoot alright. Miners only worked one shift during the daytime, and after that shift left, the shot firer would fire all these holes underground and knock the coal down for the men to load up the next day. They loaded this coal in little cars that the mules pulled out to a section and then somebody else pulled it onto the shaft. When they got the motors, this was all combined. They'd pull them on out to the bottom* and there would be another mechanism to load these cars onto the cage* and the hoisting engineer would pull them out of the mine and dump the coal. However, it was all a slow process at first.

Frederick Boch

Fred Boch at sixteen—like other young miners—accompanied his father to the mines to load coal.

I went to work like Frank Boch, my brother; I started at sixteen. Dad took me down below and said, "Well, if you are not going to school we will put you to work." I remember the first time going down; you got on the cage* and you didn't have no doors shut or anything. Whenever the cage went down, the wall went by you as you went down. You didn't dare move over to the edge or you would be scraped off there, you know. You had two bars to hang onto and you put one hand on the bar. There was sixteen men got on the cage at a time. I remember; I hung onto that bar. Usually somebody would tell the engineer, "We've got a new man going down! Give him a ride!" The engineer, boy, would what you call "cut the rope." Boy, he dropped that elevator and it felt like you're a goner. By the time the cage started slowing down, you were at the bottom. It stopped down there real easy; everybody would get off.

I hand-loaded coal. This coal was on the solid*, just like these walls here. You had to drill these holes in the coal by hand. Dad had set up a jack and augered that hole in. It cut that hole in the coal for the powder. Nighttime the shot firer came around and would shoot. They didn't shoot during the day when all the men were down in the mine 'cause sometimes the shot would blow backward. There'd be an explosion in the area or something; it was dangerous.

I remember Dad always asking me, "How much powder do you think we ought to put in this shot?" It was supposed to blast off maybe about five ton of coal, so I would say, "Oh, I'd say about a cartridge and a half." "Oh, no!" he'd say, "That is too much." But he would always ask me next time. He would always have me drill a hole and then he would ask

Figure 44. Entry clean up gang underground in Peabody Mine No. 58, mid-1920s. (Courtesy of the Boch Brothers)

Figure 45. Blacksmith shop at the Christian County Coal Company, East Franklin, Taylorville, circa 1906. The company blacksmith sharpened the miners picks, shoed the mules, and took care of much of the mining equipment. (Courtesy of the Boch Brothers)

me how much it would take. The next time I would say, "Oh, we better put in so and so." "Oh, no, that ain't enough." He would never go by my opinion, but he would always ask me.

You had to buy the powder through the company; that was the only way you could get it. You couldn't carry it, or anything else, with you. You had to have it right there and leave it there. The company delivered it, but you had to pay for it. I remember sometimes we would come in the morning and the first thing when you come in is everybody has their own room*. You have an entry* and the rooms go off of that and there would be two men in each room. You always come into your room and see how much coal your shots would produce. Some days that coal would be scattered all over and there would be too much powder in the shot, because it blew all over the room. The ceiling is dangerous. Several times we would have got covered up if we hadn't went out of the room.

This Peabody 58 Mine wasn't too deep. We didn't have what you call a man trip* to haul coal cars here on the east side. When we would get off the cage we would have to walk in. I remember it was underneath town someplace where we mined our coal. Sometimes we would have to carry a pick in. We would always carry the pick out to be sharpened and we'd pick it up on the bottom. We would come down on the cage and if we had any tools to pick up, I would be the one to carry the pick in. I remember I would go up behind him. Sometimes the roof would look bad on the entry. I would sound the top* and he would turn around and bawl me out. He said, "Don't ever do that. Some of that is ready to fall and if you touch it, it will fall."

Yes, you could see it loose hanging up there, but I wanted to see what it sounded like. That was the main man-killer them days down below—if the roof caved in. It used to kill a man about every month, I think. Oh, they used to kill a lot of men down there.

Divided Kingdom

I also went up and picked rock. They had what you call a picking table in the tipple*. The machines don't pick rock out. The machine will load up everything that is in front of it. There are certain amounts of sulfur and rock in coal. There was about thirty rock pickers up there, all young people. I picked rock for about three or four years. Then they sent me down below after a month in the shop helping the welder—helping the blacksmith.

Joe Craggs

Mines before mechanization required a great number of men to do the physical labor—work in the dirt gang, the recovery gang, and even loading coal. There was no free time since the mines were in competition with the non-union mines.

Well, the lowest paying job and probably the worst job that you could have, was the day-labor job where you worked in a dirt gang* cleaning up—loading cars. That was where they put everybody when they started out because you used a number two banjo.* A miner had a territory* in the coal mine when they had hand loading; they had lots of territory. They had to because at one time—at Mine No. 8—there was about sixteen hundred people working there and most of them were hand loaders.

The dirt gang* would load that dirt and that sulfur would get in your eyes; they'd load excess coal and lay track and that's where most of the people started. When you went down a coal mine you were initiated to number two banjo and carbide light*. You never got to stop to put carbide in your light. The boss would hand you his light till he got yours filled up so you could keep the banjo going. Oh yes, and you worked for $3.75 for an eight-hour shift. I mean you worked; you didn't horse around.

Divided Kingdom

Figure 46. Dirt gang cleaning track underground. Peabody Mine No. 58, mid-1940s. (Courtesy of the Boch Brothers)

The dirt gang boss was always the type of person that was a driver. In other words, he didn't have anyplace else to go and he wasn't going to wander off; he stayed right there to see that you had done the job.

On idle days they would have maybe twelve or fifteen men and they cleaned roadways. Well, the cars were all wooden pit cars, so they all leaked some coal as they bounced over the rails and so forth. The coal would come down and it would fill up between the tracks and on the outside of the tracks. So every so often they had what they called a road cleaning gang. And that's what you got to do on your idle days. If you got an idle day it was very rewarding, at that time; that was money from heaven. So you worked in the dirt gang cleaning road, no matter what you did during the rest of the week.

You couldn't hardly see from one car to the other at the day's end because you were loading with a number two banjo and it was dirty work. I mean, real dirty work! When you came out at nighttime your eyes burned so bad with the sulfur that was in the coal and what have you; it was something else. And you was just completely black. I mean, not just a little bit black, you were literally black.

The recovery gang* was a gang that would take whole territories and pull the rail bars out of it. In this part of the United States we used hundred-and-ninety-pound railroad iron to hold the roof up. And they would go in and recover those things and recover the track and recover props and things like that. They had nothing to do with the dirt gang. The recovery gang was separate. We laid track behind all of the machines because everything was on rails. And then nighttime the third shift would come along and deliver what they had recovered to the working sections for the next day.

John Bellaver

When work was plentiful before mechanization, miners protested conditions and treatment with their feet. John Bellaver and many other miners moved often, looking for better conditions and better pay.

My dad took me and my brother to work at the same time. See, the miners took their sons in when they got to be sixteen-years-old. But not before. Now it's got to be eighteen, but then it was sixteen. We wasn't any bigger than that.

Matt, my brother, was working with Dad. Well, then, Matt went and worked with my brother Tony, and me and Bill went in as one man, twins. So that way we got a whole turn.* The fellows said, "Well, when I went in with my dad by myself at sixteen I only get half a turn." But my brother, we got them both. My brother Bill—he was pretty lively—used to drive mules.

Oh, we done pretty good. We didn't fool around. See, you got paid by the ton and if you didn't load, you didn't make

Divided Kingdom

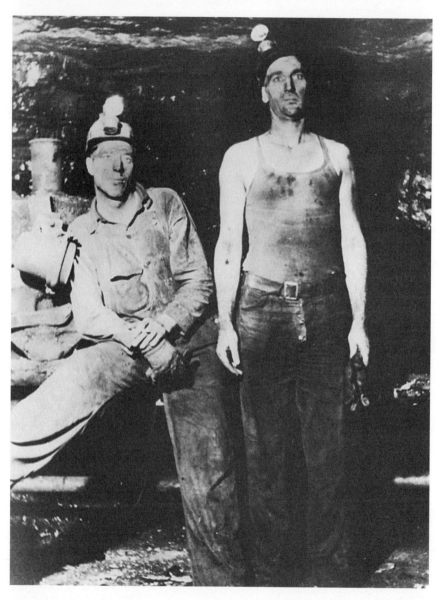

Figure 47. Coal loaders in Peabody Mine No. 58, mid-1930s. (Courtesy of the Boch Brothers)

Divided Kingdom

no money. You'd just as well go home. A good coal miner knows how to shoot the coal down so you could pick it up with a shovel.

A lot of people got hurt. That's another thing about working in the mine: laziness. You used to have to take care of your own roof. That's where they got it, in the head—from the top. If you put up a prop, it holds. If you didn't, it'd come down on top of you. A lot of miners said, "That'll hold. It's been up there for awhile. We'll go ahead and load the car." Some of them got killed.

Yes sir, you took care of your own place and your own hide, boy. If you got hurt they carried you out. That was it. And you knew what you'd done if you come out of the mine hurt. There was no ambulance or anything, brother. I've seen them haul them home in a wagon.

There were cutting machines to cut the coal. We used to cut a three-foot strip in underneath the coal. And then they'd pull it out, back it up and cut another one. They cut it thirty-foot wide on the face. Nokomis, the cutter got paid by the guy that loaded the coal. See, if we'd pick our own coal without the machine, we got $1.01 a ton. But if we paid the cutter we only got 86 cents a ton and he got the rest. If he had a good bunch of loaders, the machine man made good money. We was loading. That's all we ever done. And then when he cut the place we had to shoot that coal. I'd rather undermine it than mine it with a pick.

I was working over at the North Mine in Nokomis on pick work. I looked at that superintendent and said, "I don't think I can do this." The next day I picked up my little old Model-T Ford and I went over to Taylorville and worked there. I said, "I know where they've got machines. I'm not going to pick that coal." Oh, no. It's too hard.

You really had to have skill, though, to know how to shoot, because if you didn't, you'd ruin your top. And if you'd ruin your top, you'd get killed. Powder's just like any-

Divided Kingdom

thing else; it goes to the easiest part. And if the roof was the softest, the explosion brought it down. My dad had pretty good skills. He never drilled a hole that went up like that [at an angle], never. He always drilled straight. Said, "It'll never go up. He says, "It'll kick it down." May dad had to pick it in Pennsylvania. They had no machines there. No machines at all. They picked everything. But the coal in Pennsylvania is different than here. There, you could take a pick and bust the seam. But here the seam lays flat.

You know, I went from here [Schram City] to Nokomis to work for Coalton. And see, if you didn't clean your coal and load it, they'd dock you for it. It'd cost you fifty cents. If you got three of them, you got fired. We used to take a sledge and bust it into pieces. We got by with it. Because you couldn't clean all that stuff. I was trying to clean it and a guy came by and he said, "Hey, if you're going to do that in this mine, you're going to get no tonnage. You're getting paid by the ton. Load it. Forget it."

The company hired the machine man to load coal later in the 1930s and put him on company wage and not tonnage. See, the machine belonged to the company. Other company men were guys that laid tracks, guys that throwed rock away, and mule drivers. All them were company. Only guys that wasn't company men in the 1930s was the loader. The undercutter gets paid by the ton by the guy that loaded it, out of the loader's wages. If he had a bunch of lazy skunks working for him the machine men would get rid of some of those birds or the superintendent got rid of them. He'd say, "They didn't sink these mines just for you to come down and take a breath and then go home. They sunk them to get the coal out."

We had unions whenever I started to get hours. When I started mining, day workers were getting $4.25 a day. See, the only guy that didn't get paid by the company was the loader, and he got paid by the ton. Track layers got $4.25. Mule drivers, the same, everything.

It was about 1935 when the loading machines started coming in the mines. They didn't need no man loading it with a shovel when the loader come in. So we had to go on a division of work where you work one day and the other guy works one. You stay home one day and they work one. That's the way they worked it. If the mine worked four days a week, you got two, see. The United Mine Workers wanted to let men go and the Progressives wouldn't allow it. We were Progressives.

Alvin Amuel Wise

Mr. Wise's interview gives a good idea of how diverse the coal miner's job was before mechanization reduced the job to a few simple tasks.

When I entered the mine I was short from May to October of being sixteen. This was 1929. We were still hand-loading, see; there were thirteen hundred men working here at Peabody No. 8, all hand-loading. The cutting machine would come in and cut your place. Your room off of the entry was supposed to be thirty feet, but sometimes you can get thirty-two feet wide. The first night, if the cutting machine would come in, you'd drill your holes half way up. Say the room was seven feet tall—the coal vein was seven feet tall—you'd come up three and a half feet or just under three and a half feet and you drill your one, two, three holes there. You drill two snubbers*, they called it, one on each rib* of coal. Then you cut your fuse so that the middle shot would go off first; that would break the middle up. Then you shot the ribs which would break. So the cutting machine cut ten or eleven feet deep; you would only have a bottom part to load on the next day, unless you had two rooms. If you had two rooms then you would have extra coal, so the next night you would drill the top. Then you shot the middle and the two sides—the two ribs. That would break it down and the next day you would have plenty of coal to load for one day.

We were using black powder. Only if you were shooting rock or something like that, you would use dynamite. But coal you always shot with black powder. Dynamite has that boom, you see. It goes off all at once. Black powder would spread this way and up and down. Dynamite, most generally, all the pressure goes down. But you have to be careful where you put it; you got to put it where you think it will blow.

It wasn't hard questions they asked you when you wanted to be a coal miner. But at that time, they asked mostly about the face and how careful you were. They asked if you timbered your room, if it was bad slate or rock top; what you'd do; and if you'd put up your timbering with steel bars or wood posts. They didn't exactly call them posts; they were just round oak legs. You put them on each side of the track. You always had a track running up near the face at that time, so you could get your car in there to load it up.

You timbered on each side of the track up to a certain point. You couldn't do it right up the face. Say you'd come to within fifteen feet of the face because you had to have room to work up there. After you got everything all cleaned up then you could go and set timbers or whatever. They didn't have roof bolting at that time; all they had was steel bars. They had some places that would have wood bars and legging.

Well, I know coal mining; that is all I ever did (laughter). I know coal mining in and out even today. I have worked on every piece of machinery there is in a coal mine. I just learned. They never really taught me. They would say, "There it is; it is yours; run it."

Donald VanHooser

Donald VanHooser describes what it was like to work in a mine undergoing mechanization, and what machinery was being introduced. VanHooser had enough foresight to realize that without sufficient technological experience, he would be without a job when the transition to mechanized mines was completed.

My dad was a coal miner. He first worked at Mine 58, here in Taylorville and then he went to No. 7 at Kincaid. He was there several years while I was in school. At Mine 58 he was what they called a shot firer. He had to go to work in the afternoon and do the firing at night, after the mine closed. When he went to Kincaid, they had loading machines; here at 58 they didn't.

When he went out there to Kincaid, cutting was all gang work—no miners. After awhile, he got to where he was running a cutting machine for the gang. Of course, you had to have help. To begin with, there was three men in a gang, and when the mine got big enough they developed other territories and they added men until they got seven-men gangs.

The men on the territory loaded in three-ton carts. There was sixteen rooms to a territory to load—sixteen rooms on this side for one territory, and sixteen on the other side for the other territory. The air went in one entry and came out the other, and that was your circulation, see. The cutting machine cut on both entries. Of course, you had the helper on the machine.

They had what they called a dock boss. Well, two cars dumped in the hoppers up in the tipple*. The cage* would hoist two cars at a time. They had two cages, one would go up and the other come down, and there's four hoppers. Of course, the diggers had their own check number, a little pasteboard check number on the cars on a nail that was

Figure 48. A cutting crew maneuvers a Morgan Gardner cutting machine into place at the face. (Courtesy of anonymous Peabody mner. Donated to Illinois State Historical Library, 1988)

bent over. The cager in the tipple, he would dump one of those hoppers at a time. He worked pretty fast: dump one, walk over and dump the other. But the coal was separated just slightly, and he had a phone right there. He'd get one with a lot of rock with it; he's stop the tables and call the tipple and get the check number. We'd throw it out if there was just so much. Why, it was called a common dock. They fined them, I think, fifty cents. If there was a whole lot, it was an aggravated dock and they were fired. Those diggers could tell when they were loading the coal by hand; they could tell how heavy their scoop was whether it had a chunk of rock in it or not, if they wanted to.

On my first day at 58, all we did was load coal. It was five or six cars that we loaded, and they were three-ton cars. You had to take your turn, as they called it, whenever the motor would bring in empties, and take back the loaded cars. He would push your entry on past the room and go in and get your loaded car and take it out. You'd push your empty car into your place. When they would gather all the coal cars on that entry they would take them to the intersection out there on the main line. The main line motorman would bring in empties and take away the load back to the bottom. They were hoisted from there.

The first day in, I tell you, I was sore. I was sore and stiff, oh! But the second day, we drilled three holes up there to shoot the place down because it was cut underneath. First, of course, came the cutting machine. The cutter bar was six foot long, and you had a bit chain—pig chain, it was called—and it was fed through a sprocket on the back end of the machine and around over here to the ratchet and up front to another sprocket up there, then across the room thirty feet over here and you anchored it in a wall or to the top of either one. It was already cut and we drilled it.

Then Dad made the powder up and showed me how to make a cartridge. The cartridge was a big dowel rod, if you want to call it that. It was made for that purpose, because it was tapered. You used newspapers to make your cartridge. You kept wet clay and you put your cartridge in there. You measured so much pellet powder, and you put that in your hole. If you had three cartridges, the last one had the fuse in it. You put them in there and then you put clay in there and tamped it tight so that your core would stay tight.

The second day is when Dad stayed in and shot. You had your fuse long enough that you had a good fifteen minutes to get away from there. So he had me out on the entry with our dinner buckets to come home, and he stayed in there and shot. He didn't shoot until a quarter of four or something like that. Anyway, I tell you, the first boom blowed my light out and there I was in the dark. I was really scared. But Dad came to me right away and we came on home. But I tell

Divided Kingdom

you, that was dark! But I soon got used to it. We worked those rooms for over a year.

At that time you had Morgan Gardner cutting machines made in Columbus, Ohio. The machine itself, as it crossed the front, was forty-two inches across. There was a big electric motor and it ran the cutter chain, and that chain went around the bottom, cutting. It was on rack rails, they called it. That machine fit in on the rails and then backed in and out; then you moved it over and cut the face again. You made four of them in a twelve foot entry. My job, at that time, was to set the front jack pipe, that was a little short one you set in the coal. But you had to know how to chip that out so that it would fit in there tight. Dad would set the back jack. We got along alright; but boy, that was working. Many, many, evenings I would come home and just lay down on the bed dead tired.

The machine itself, when you moved it from one room to another, went on a cart and you pulled it. You could move it from one room to another by itself. You had a big, long electric cable. Dad one time cut his big toenail off; it just caught on the machine and come off his foot into his sock. Well, he was off several days with that. There was no one to run the machine in the gang. But Dick Klemm helped me and we cut coal. I ran the machine. Some of the older men didn't like a kid eighteen years old running the machine when they were just helpers, see.

When was it that the loading machines started to come into Peabody Coal? That was beginning in 1925. I was twenty-five years old when they were coming in. I told my buddy, Chuck Simmons, "Well, if we're going to work in the mines, we'd better get on a machine of some kind, because they'll be laying men off."

Later I made an application, or said I wanted to get on the loading machine, and they put me as a helper and I learned to run the machine. That was what they called a little 5BU*. When the next number five machine came, I was capable and I got that machine on the territory. We had those machines several years. I'm going to say several hundred men were laid off in that one mine.

Joe Craggs

Peabody Coal implemented major technological changes in coal mining in the 1920s and 1930s. Such changes necessitated a reorganization of the work assignments. Joe Craggs details the changes in the hierarchy of work, the nature of innovations, and the work processes.

The Peabodys had grown up with the United Mine Workers and John L. Lewis. John L. Lewis was—although he was a powerful labor representative—a fair man as far as union and mechanization was concerned. He did not believe that you could not mechanize the coal mine. That was what a lot of the unrest in the unions was born from. Mechanization was coming about—becoming of age—in the coal mines. Up until that time you had hand loading which was, oh, just pick and shovel work.

The coal-mining industry started in Europe where it was all hand mining. You did not have a machine to do anything. You shaved the shot on the solid* or you made your cut with a pick. That is the reason a lot of them came over here from England and were experts at using a pick. On the solid—the only mine that did that was Mine 58. And that was many years ago that they shot on the solid and they put in what they called lifter shots*. The lifter shots would go off and then that would make their cut for the other shots to go off.

My first coal mining was done with an old punching machine, they called it. A punching machine was a cutting machine that went in and out. That made the crust crumble under the coal, or in the coal, so that the coal could be broken both ways. There were machines that you could take a tow rope or chain and pull across the face to cut the whole place instead of punching it in and out. They were usually manufactured by Jeffrey Manufacturing Company of Chicago [1919-1938]. Jeffrey and Goodman were the first cutting machines that we had. That's how they came about. Then your next cutting machine was on track. And it was a cutting machine that swung around, the bar swung, and it stayed on the track. And then your next cutting machine was on cats [treads].

The first drilling done in the coal mine was done by hand. They used what they called a breast auger* and it was a harness that went around a man's chest. He pushed on it literally with his body and turned it and he drilled the hole. That breast auger, that is how he shot his coal down. But he tamped it with black powder and newspapers.

We were drilling with the breast augers; then they started with post drills. The post, that was before the cutter; so it had to be probably in 1916 or 1918. By that, I mean they set the auger up on the post and it had a box—a thread bar—and they would turn this, and this thread bar would feed this into the coal. It was a number of threads per inch that caused this to go into the coal and determined how fast it went into the coal and so forth, how much pressure you would put on it. It was a brass box that they called a thread bar. They still use this today in a lot of coal mines—this thread bar. It clamped whenever you wanted to release it; you just threw the lever back and it slid through the thread bar back. That was a post drill.

Anyway, the drill bit that they developed from that was a cat paw that came into effect then too. But there was an expansive-type drill bit that would go in the hole, through the hollow auger to the center. They would then force this drill open on the inside of the hole and they could ream out the pocket in the back of the hole. That way they could get more black powder in the back of the hole, bring the hole down and shoot it out—loosen it up.

They used that for quite awhile—the hand loaders did—until the development of higher strength powder and so forth came about. Then your next drill that came about much later was a Dooley drill—a semi-portable—which you set up on your post. It was an electric drill. You set up the post and then you picked the drill up and hung it on the post. And you swiveled it around and drilled your holes. That was the first electric motor which was all DC current. These drills started out small, but they got bigger and the company made the cases out of aluminum that was mostly developed by Dooley Manufacturing Company in Peoria. Then from there we went to rubber and we built the first ones right here in Taylorville, right out here at Manson Machine Company. It was designed by Mr. McCann, myself, and Jimmy Young. We designed the turn tables and everything and made the whole set-up. In the late 1960s or early 1970s, they came in with one in which a man could set back here and operate two big booms hydraulically. And the bits were forced into the coal hydraulically. Now they don't even use drills. Now they use the continuous miner* to take the coal out.

At first they were shooting black powder, which did not have the umph that powder did in later days and was developed by DuPont and so forth. Due to development of black powder they couldn't get enough powder in the hole. The high power explosives hadn't been developed yet.

At that time the coal company and the consumer wanted lump coal, anyway. The coal miner was paid for lump coal. The hand loader was paid more for the lump coal and if he didn't put so much lump coal in there, they would dock him. I mean, they knew; they didn't want fines. They wanted lump coal, because lump coal was a household word; you got a big chunk of coal, then you would take it out in the shed and

Divided Kingdom

you would beat it up with a hammer before you would put it in the stove. But you couldn't pick it up. But they wanted lump coal. They thought that was the cleanest, which it wasn't. The only thing that held lump coal together was rock, dirt. That held it put.

In those days we did not have roof bolts so the danger was very real. In other words, right up to the top of the coal there was about a four-or five-inch seam that was just like mud and when the air hit it, it would come down in slabs. More people were injured or killed in the Midland coal mines here by the clod than anything else. The air would hit it; boom, it would come down. When it fell it would smash them flat, or break a foot or leg, or something like that. And the clod man would come along, ahead of the loading machine. He would take down everything that was loose. He had a wedge and a sledge, and he would wedge and hammer and pick at a bar and would pull down this loose material. The top above that was what we call lime rock top, which was very good top as a rule. A lot of these mines here in the Midland Tract had that. That was one reason it was good coal mines. And they would clod the place and then the machine would come in and load it up and it was a continual process.

In those days we didn't shoot on the same shift. You couldn't shoot on shift in the State of Illinois, like they do in Kentucky, in Virginia, and so forth. Here you had to shoot in the end of each shift. So whatever the loading machine loaded—say, if he loaded up fifteen places a day—you had to have that many places ready so the loading machine would have the coal ahead of him. If you didn't have the coal blown down, why, you'd run out of coal and you got fired if you was the face boss.

We would bring the drills and machines to this area, which was known for its high production per units of coal. The manufacturers were very eager to get their new machinery into Peabody's coal mine because of the high production of the people. The workers here at that time had the idea that the employers didn't owe us a living. We had to work for a living. Therefore, we produced for Peabody Coal Company some of the highest tonnages that has ever been in this world. Tons per man. That is true. I have records to prove it.

The new Mine No. 10, sunk in 1951, probably is one of the most fabulous mines that has ever been built in the world. But they have, say, maybe five or six shafts where they would bring the coal to one place and they would call that a coal mine. That isn't what No. 10 was designed for. We produced over five million tons annually for about fourteen consecutive years in No. 10. Which is better than any coal mine has ever done in the history of the world. With the number of people we had, the tons per man used to go all the way up to thirty and thirty-five tons per man. These people were workers. Wonderful workers. And they were people that were proud of their heritage and proud to be coal miners. If you look at the communities of Taylorville and Kincaid and Bulpitt, the homes are well-kept, and they were just proud people. Most of them in Mine No. 10 were the second generation of coal miners from the old country.

At that time, in 1929, when we began to introduce machines, there was no seniority or anything else. It happened through attrition more than anything else, if you understand what I mean by that word. In other words, if you were mechanically inclined, you would make up a coal crew. They did not mechanize the whole coal mine at one time. They kept the hand loaders in an area, and the hand loaders would still work, but the machines would be in another area, and finally through attrition we were able to replace men with machines. And that's what the Progressives couldn't understand. John L. Lewis, in one of his famous talks about mechanization, made the statement that we had to mechanize. "Progress, you cannot top progress." And he

Divided Kingdom

believed this, whole-heartedly. That was the reason we mechanized so easily in this area. But we did not start it all at one time.

The first loading machine that was developed was called the Joy Junior. The Joy Junior was made from Model A Ford parts. We used to buy a lot of our parts for these first loaders from Henry Ford. And it was made by a man by the name of George Joy. He was originally from Centralia, Illinois. Then the Joy Manufacturing Company bought those rights out from him. Then the next machine that came along was the 5BU.

The younger people were the operators of the loading machines. The older people were the drillers and cutting machine people. See, on a crew at that time you may have up to twenty-four and twenty-five people. When we first started mechanization, though, we had up to thirty because it was all track loading. Your crew consisted of a loading machine operator and the second and third man. Then, as a rule, you had a motor man and a trip rider service them. Then, on top of that, you usually had two sets of drillers. One was a driller and one was a shooter. They tamped their own shots and got them ready to shoot right after quitting time. The drifters always came down on what we called the nine o'clock cage. They always stopped the hoist at these hoisting coal mines and let the drillers and shooters down. They stayed after quitting time. That was when we were single shifting. You got to remember, this all started as single shifting coal mines. They may have had a couple of machines on the second shift to fill up the empty pit cars that were put in storage for the next day, but they did not hoist coal until the 1960s or 1950s on the second shift.

So they had operator drillers, they had track layers, they had switch layers, and what they call straight track layers who kept the track straight so the cutters could get up to cut the coal. The track had to be laid up to the facing. Then it had to be recovered. Then we had a recovery man go around to recover the rails that were up to the face when we were done. The track would lay right up to the face so the machine men could sink in and make their cut. Then when they pulled out, there had to be a man come around and recover the jumper rails, as we called them at that time. So when the coal was shot down, the loading machine would go off the end of the track, and he would not get hung up on the track, see.

So, they recovered that track back maybe twenty feet from the face and he would load into the pit cars. He would load into the three-ton mine cars. Well, at first they were about three ton and then they put eight-inch doors on them in this country. The second and the third man had to get them eight-inch doors up as they came in so that the loading machine could load the cars higher.

Your categories of paying jobs in the 1930s was first your day labor—which was anything as a clod man, a clean up man working in a dirt gang, recovery gang, or menial jobs. You would go on up from there. Starting in the thirties as a motorman and a trip rider, in those days. That was in the pay scale after the clean up man and so forth. Your trip rider was paid forty cents less than the motorman. Then you went on up to your third man, and your second man, and your loading machine operator. Your driller was paid about the second man's wages. Your cutting machine men were paid the same rate as the operator. They cataloged in that fashion. Then you always had track layers and their helpers. Your track layers were paid twenty-five cents more a day than the track layer helper which was a menial pay scale of $3.75 in the thirties. And then it went to $4.25; we had a fifty-cent raise.

Then you had timbermen who were paid the same scale as the track layers. Oh, you have different road motor men. You had all the bottom crew which were about the same scale as the track layer.

Divided Kingdom

Well, a mine boss was paid practically the same as a loading machine operator. I mean, very little difference in the scale. There would be about four sections in one territory for loading machine crews. Then they usually had a monthly man in that area who was the head boss. He was paid a monthly salary. When I started, the monthly salary was $400 a month, which was lots of money.

The mine boss usually worked one day a week more than the loading machine operators or anybody else. Consequently, he made more money. In them days, you wanted an extra day's work; it was pretty hard when you were bringing home $6. When I started face bossing in the 1930s, it was $6.40 a day. I still got some of the statements left. My wife saved everything.

Table I.
Inside Day Wage Scale

Central Fields, 1922

Tracklayers	$7.50
Track helpers	$7.25
Trappers	$4.00
Bottom Cagers	$7.50
Trip Riders	$7.40
Motormen	$7.90
Timberman	$7.50
Driller	$7.75
Cutting Machine Operator	$7.90
Mine Face Boss	$7.90
Head Underground Boss	$400.00/mo.

Source: Isador Lubin, *Miners' Wages and The Cost of Coal* (London: McGraw Hill Book Co., 1924) p. 207.

John Ralph Sexson

John Sexson was a pro-Lewis man during the Mine Wars. He documents the effect mechanization had on miners, especially the reassignment of the work force to new jobs.

First, in the 1930s, they changed over to what they called loading machines. The coal would be drilled and shot. Then the loading machines would go in and rake it up on a big arm onto a conveyor and then back into a mine car, what we call pit cars. They would hold about three tons each. When the motorman got strained—what they couldn't handle, twenty to twenty-five of these cars—then he would go to the bottom and they would be dumped into what you call a shaft. At that time, when these loading machines came in, the miners raised quite a hell about it. But John L. Lewis said, "You're going to have to do this or else we're going to be out of business."

Mine No. 8 employed twelve hundred men, loading by hand. When the loading machines came in, that cut off five to six hundred men. They would go out there for their turn at working and it got to be where you just simply couldn't do it, so they lost their jobs. But if they hadn't done that, the mines would have been down years ago, I believe. They had to come to a way they could get more coal out and do it cheaper.

So with loading machines, I think John L. Lewis was very smart in allowing them, instead of calling a big strike to try to stop them. That would have been impossible. And that is just about the time the Progressive Organization was in. They didn't like those kind of things where they was almost dictated as to what to do. John L. Lewis, although he might have been a dictator, was a very shrewd man. I know they formed the Progressive Union and tried to break up the United Mine Workers, but it didn't work. So we had a little battle of a year-and-a-half, two years.

Divided Kingdom

In 1931, they started putting the machines in. Peabody Coal Company, at that time, was up on the most modern mine machinery you could get. And you couldn't believe it. When I worked on top and watched them hoist this coal and dump it, I had no idea. I couldn't understand how they could get those cars on so fast at the bottom. You see, they dumped four cars in about fifty-six seconds from the top to the bottom back down.

Frank Boch

Frank Boch relates another effect of mechanization—the displacement of older men by more adept and unskilled younger men in the UMW mines.

The way you got a job in those days was if you had a father working in the mine. You had to work with some experienced miner that had mining papers for two years so you could get some papers. So I went to work with my dad. The first job was hand loading. You had a room down below and you blasted the coal with powder [shot the coal]. The next day you loaded the coal into the cars and then you did the same thing over again every day. Blast the coal out for loading the next day.

In five years after I started, the company decided to mechanize the mines and brought machines in. I was a young fellow about twenty years old, and I got a job hauling coal. I was what you call a trip rider. I helped couple the cars together to take to the bottom. We were in Mine 58 at that time. I would say there was about eleven or twelve hundred men when they had hand loading in 1929. When the machines were established I don't think there was about three hundred.

I guess they could lay off anybody they wanted to. Things was in my favor because I was a young kid, not like an older miner who couldn't do that kind of work around machines. You had to be kind of active and in good physical condition to do them things.

I remember my dad couldn't have been very old when I went to work in the mine with him—up in the forties, or maybe close to fifty. He had what they called miner's asthma. They didn't call it black lung like they do today. In fact, they didn't know anything about the danger of that dust and stuff. I remember he couldn't breathe and he'd just have to hang on to catch his breath. He was like I am today, only he was a lot younger.

He wouldn't have been fit to work in that new mechanized mine because you had to be pretty light to be around them machines and stuff.

He retired early. There wasn't no benefits back in those days, and he was lucky that he joined the United States Army where he got a small pension. He didn't have much money, but they had enough to get by on. Then he puttered around here and made some concrete pottery and stuff like that. He liked to do that kind of work. He made a few dollars on that, just side work.

John Bellaver

In the PMA mines readjustment to mechanization and the cutback of labor took the traditional form of time share. John Bellaver knew what was coming to the PMA mines, though: shutdown.

With machines you needed two drillers and you needed men to prop the roof so the roof doesn't fall. And you had to have a cutting machine to cut the coal so they could shoot it. I think there was ten or twelve men used to follow the cutter. More than that. Also,

Divided Kingdom

Figure 49. An out-of-breath William Augustus Pierce poses with his Model-T Ford in 1923. By the time most miners were in their late forties, they had chronic "miner's asthma." Pierce died in July of 1923, and had rarely worked a "full day since he turned 60." (Courtesy of the Boch Brothers)

one loader. Now with the miners they can get by with about six men.

When they first started using these loading machines, they used it down here. Nokomis started it. It was about 1927, 1928, something like that. It threw a lot of men out of work.

Loading machines was already established [in Nokomis] when they come in other places, see. The Progressives tried to share after they started putting in the loading machine. Then they started putting in the continuous miner loading machine and it was worse yet. They don't need no men at all. Only four or six men. The mines worked every day, but not the miners; it didn't take no men. They just laid them off. Said they didn't need them.

I worked under this loading machine in Taylorville up here for five years. Them that went back, went back as day work. They didn't need you a heck of a lot. There was no work. All the mines were mechanizing. In the 1930s was when they really started out with the loading machine. That's when a lot of our relatives went to Kenosha to work in the factories. I got married in 1929 and Tony, my brother, had just moved out and went to Kenosha because I took the house that he made. And my folks went to Chicago.

We did share time when the loading machine came in under the Progressives; but under the United Mine Workers, nothing doing, no share. Under the PMA, we said, "Hell, you ain't going to throw us all out. I'm just as good as you are. I had a job here, too." You know what the United Mine Workers did? Other mines took the oldest man, and he stayed. But they got around that. They'd fire a guy if he didn't do this work. They'd put you on a job they knew you couldn't do, see. Well, then they'd fire you. The miners that was smart knowed what was coming [shut down].

Company Practices and Safety

Underground miners were aware of the precariousness of their lives. Tales of death and disaster comprise a formidable collection of first-hand accounts in the interviews we conducted. Death and injury came from coal and rock falls from the roof, gas and coal dust explosions, and haulage mishaps. Between 1906 and 1935, 47,828 Illinois mine workers were killed; there were 1,594 fatalities a year, and ten thousand more injuries.

Accidents increased as the mines mechanized during the 1920s. There was a shift from mules to motorized haulage, and new, inexperienced men were hired to run the new machinery. Also, the deskilling of the miners' traditional jobs was a factor, too. Roof falls, mostly due to inexperienced men, remained the primary cause of death and injury throughout the period, accounting for 53.1 percent of all coal mine fatalities in Illinois between 1906 and 1935.

The coal companies skirted responsibility by placing the blame for accidents directly on the miners. Management generally assumed all mine accidents could be traced to an individual error on the part of a miner. Whether conscious or not, this deception relieved the employer of all moral and financial responsibility for accidents. An investigative report in 1919, for example, declared that a "large percentage of the fatal accidents may be attributed to carelessness on the part of the person killed."

There was a variety of reasons why an increase occurred in the number of accidents in the 1920s. In addition to inexperienced men, there was an overall speed-up caused by heavy competitive pressure. The rise in productivity was directly linked to the cost of life and limb to the miner.

Competition from non-unionized fields (the initial incentive toward the introduction of machinery) placed great pressure on Peabody Coal and other companies to get as much production out of their men as possible. One particularly effective method was Peabody's pressure on the face and shift bosses to meet tonnage quotas. Another more traditional means was to encourage competition between loading crews. The most coercive tool was to threaten a miner's earnings by placing him in a dangerous and unproductive workplace, or by demoting him to the dirt gang. The company made it clear that it would impose its will upon the miner and speed up production no matter what. Whatever the means of pressure, the effect was the same: an increased number of accidents caused by roof falls and haulage speed-ups.

The competitive pressure and the disregard for safety were confirmed by the findings of the U.S. Coal Commission in 1923. The Commission reported that "disregard for safety is caused by continued pressure on coal mine superintendents to increase his output regardless of dangerous conditions." Other confirmation of company pressure in the Illinois coal fields could be found in the high labor turnover in the mines, the increased rates of absenteeism, the intensified number of mine accident investigations, and the number of days mines were closed due to unsafe conditions.

Heavy competitive pressure was also a reason that unsafe mines were developed, even though the engineering expertise was available to make them safe. The 1923 U.S.

Coal Commission Report also found Illinois' larger mines, including Peabody's, were "developed in such a manner as to bring about the squeezing and crushing of pillars, and the heaving of bottoms, and the disturbance of the roof with increased hazard of roof falls." The main objective of the owners was to establish mines that were most accessible and the least expensive to develop, leaving aside questions of safety. The 1923 report continued: "such mining practices [as undermining the pillars] ruins the roof, and when recovery teams go in to extract coal from the pillars, they find coal broken and the roof treacherous."

The Illinois coal operators blocked the attempts by reformers to remedy the problems caused by speed-ups and unsafe conditions. In 1929, a simple bill was introduced in the legislature requiring industry to remove rock dust from the main haulage entries and rooms to within forty feet of the face. The effect of the bill was to reduce explosive conditions in the mines, but the operators defeated the bill with ease. The inability of government to ensure some degree of safety was predictable and resulted in a dependence upon one common denominator—the speed of the boss.

Duke Allison

Peabody management encouraged competition between work crews, and punished and rewarded miners by awarding desirable work locations. Jobs were difficult to obtain during the 1930s, so miners such as Duke Allison stayed and tolerated the conditions.

By working and observing you got better positions. Sometimes a guy would be laid off and they would be short a man; then they would put you on that job. If you did that job, why then the first chance you got you would get a promotion. Seniority didn't work at all. The management ran the place. They could take you off a high-paying job and put you on down below, or they could take you out of the dirt gang and put you on a loading machine or cutting machine. You got ahead if you knew somebody. Just like politics—if you knew a politician you could get things done. That was the way it was out there. If you was a cousin to the superintendent, why then you wound up with a fairly good job.

My dad was working at that time [the 1920s]. This is the way the younger guys got started in the coal mines if their dad was working there, and if their dad was a good worker they would being the boys out. There was families on top of families out there because the dad got them all in, five and six boys.

There was competition between the superintendents of the mines because everybody was trying to get more tonnage and everything. But as far as the men were concerned, you worked in what you call a coal crew. At that time they took pride in their work because they wanted to load more coal. They used to come out at quitting time and they said, "Well, we loaded two hundred and twenty cars." And here is another guy from another crew who said, "We loaded two-forty." At the same time you had to have a little pride in your work, and you had to do your work or otherwise you didn't have a job.

I experienced layoffs, like when No. 8 closed down every April. They shut down for thirty days. Now, if you had the connections, you could be given a transfer to one of the other mines. Well, at that time, I had a brother-in-law that was pit committeeman who represented all the union men at 58. So, I got a transfer to 58.

They shut down the whole operation of No. 8 for thirty days. Then the next month, No. 7 would shut down. The next month No. 9 would shut down. So you was going to be shut down one month every summer. But 58 didn't shut down;

Divided Kingdom

they worked that year 'round because they had a contract with Wabash Railroad to supply coal for the locomotives. They had a different contract than the other mines had. Our coal was going to Chicago. It was five days a week. I made more money in that month than I was making two months at No. 8. The mine just operated two or three days a week.

Jesse Lake

Jesse Lake, a UMW loyalist, relates how management made work difficult for members of the "dirt gang" and young boys on the picking tables. The miners were also subject to company money-making schemes such as buying company insurance in 1935 and 1936.

The worst job down below used to be the dirt gang. So if you did something that the company didn't like, or if you took up a case that they didn't like, they would shaft you, so to speak; they'd put you in the dirt gang. Which I guess was a dirty job loading old, rotten dust with a shovel by hand. That's where they put troublemakers. But a lot of the cases were people standing up for their own rights. Of course, back then in the 1920s they always said there was twenty guys out at the gate looking for a job; so, if you didn't knuckle down, so to speak, you were out.

When we first started they used to have picking tables where the coal would go along and you'd pick the rock out of it. They had boards where you could sit down a lot of times. Later, they got tough and felt they better get rid of those boards, because they thought people sitting down on their rear wasn't doing enough work. There would be two to four men who walked on these conveyors, and that was hard work, bending over and picking that rock. Trying to walk on the coal and bend over a lot of times—that was pretty tough work. Of course, when you're sixteen, eighteen years old, you can work a lot harder.

And if you're on top, they would probably put you to picking rock. Of course, at the time, picking rock was a cheaper scale than the yard work. I think there was a dollar difference in the day's work, and everybody would strive to get out in the yard.

I haven't got anything against the company because I made a good living later on. But I can remember years ago, they did things out there I know was illegal, such as stickers. These stickers was a means whereby if you were short of money, you would go to the office before you went to work that morning and tell the people in the office you wanted to get, say, a ten-dollar sticker, or a twenty-dollar sticker. They would charge you ten percent interest on that damn stuff. The sticker actually was a loan. I don't know where the term sticker come from, other than they really stuck it to you. So maybe that's where it come from, I don't know. At payday, you had ten dollars checked off of your paycheck, but actually you only received nine, which they were charging you a ten percent interest on that money. To me, I always thought that it was illegal as hell. I know that it was about five or six times more than the going rate of interest at a bank. I don't know who got the interest; I guess, the company. But once you got hooked on that, a lot of times you didn't draw hardly any paycheck at all.

Talk about company harassment! I remember one time they were selling insurance [1934, 1935] and this friend of mine, Merle Ahlberg, and I were the two remaining people that didn't take this insurance they were trying to sell. It was company insurance and it didn't pay very much hospitalization. If you got hurt at the coal mine, for compensation someone would pay for your hospitalization. But anyway, I wasn't making much money and they wanted everybody to take this lousy insurance. So finally the tipple boss came by and told us that the superintendent wanted to see us at the office. "What about?" we asked him, and he said, "Well, about that insurance." So Merle and I proceeded to go up to the office. The mine superintendent asked us why we didn't take that insurance, and we said, "Well, we didn't want it. Why, do we have to take it?" He says, "You don't have to, but I suggest that you do." So we signed up for the damned insurance that we didn't want. That's what I'm saying; they put pressure on you for stuff that you really didn't need. That's what bothered me so. I knew it didn't pay very damn much—the benefits—compared to what you had to pay. I can't remember how much it was and they checked it all off of your paychecks. But I ended up taking it. Probably if I hadn't they would have either harassed us until we either quit or took the insurance.

There was quite a change after the Second World War because it seemed like everybody had grown up, if you want to call it that. When you came back, they didn't try to shaft you or give you a dirty job or a lousy, hard job, because people wouldn't put up with it. You know, you've been in service and you saw people killed and stuff like that, and your outlook on life was different. You'd say, "Well, I'm not going to take no more of that crap." I know I had a different outlook on life. Of course, I had a brother killed in World War II and I didn't want anybody giving me any crap. That's just what it amounted to. I think I would've fought at the drop of a hat if they tried to shaft me.

The rules hadn't changed, but the attitude of the management had because there was a lot more work and everybody had plenty of employment. I'm talking about 1946. Everything was booming then, and they're wanting to get back in production. You could get a job anywhere, regardless if you was a coal miner or working in a mill or anyplace—you could get a job. So they didn't have people looking for all this work; back in the 1930s, everybody was looking for work. There was a difference in the attitude of the company towards the miners and the union.

Cuthbert Lambert

Cuthbert Lambert's history details the speed up; the placement of unskilled and untrained miners in skilled positions; the use of old, worn-out machinery; and the disregard the company had towards all kinds of safety issues in the 1920s and 1930s.

In the late 1920s, the pit committeeman system was a joke. They really didn't think they had much to do, and really weren't getting the job done like they ought to. That's why they eventually went to the three-man committees. They figured you can buy one man, but you can't buy three. There's always one in that three that is going to squeal. It is just like raising kids, 'cause they will buffalo you. You raise two and then the two of them will gang up against you; but if you raised three, that third one is not gonna go along with the other two. There is gonna be one in the pot at all times that will buck, and that's the same way they figured out there.

Sometimes the old fella who was committee rep would try, and other times he was a little scatterbrained. The guys got to thinking that he was being told what to do and what not to do, and letting things slide. I can't say too much, because four years there I wasn't at work; but before that, he wasn't too sharp. At least the men thought the company was telling him what to do, instead of him telling the company what to do. He was the go-between to the company on the cases. They got away with a lot of things they shouldn't have.

The company would naturally try to buy the committee, 'cause they wanted to run that mine to suit themselves.

Divided Kingdom

They even tried to carry men out of there—the mine—they knew was going to die—get them out fast—so they wouldn't have to shut down. There was a time or two, they claimed, they were alive when they brought them out, when we knew good and well that they couldn't have been because they had a lot of injuries.

They were pushing too hard in the thirties and trying to get too much done with too few men. I would say then, at that time, we worked and had probably 615 men—union men—on the payroll. We basically got out a hundred thousand ton a month if I remember right; they were looking for that four, five million a year or something like that.

When I wasn't drilling, I was helping the shooters or running the extra loading machine in the gang. If they got enough places of coal ahead they would load two machines, because they had the continuous belt running. They'd put an extra buggy* in there, and the extra loading machine, and if you got enough coal ahead that boss had to get somebody else on that other machine, if he only had one man on it. That's what was bad; they would put on a man that wasn't qualified with him, and it took an operator, a helper, and a buggy runner—three men—and only one man was there. They would get two hundred buggies a day out of a section, easy, and they could get it shot and drilled.

They just tried to take that old machinery and rebuild it and see how much they could get out of it without buying new stuff, which eventually they had to do, anyway. So it was just a push for production. If it runs, run it. Hell, I have seen them run a cutting machine down there when they had two men carrying oil to put in it to keep it going. All they had to do was shut it down for a couple of hours to change the seal on it. They would pour three or four barrels a shift through it just to keep the damn thing running. That was the Peabody system; if it will move, run the son-of-a-bitch. We can get another tomorrow. That was their theory. Well,

Divided Kingdom

Figure 50. "Typical" coal car accident on the Chicago and Illinois Midland Railroad. Fourth Street underpass. Pawnee, 1908. Speed and neglect of maintenance caused most accidents. (Courtesy of the Boch Brothers)

see, the boss was on production, and if he could keep it going through his shift, then it was the next shift guy's problem. If it went down, then he was stuck; he had to have it fixed. If he didn't have a cutting machine, he would have to get one brought in. Then he was without production on his record, and they chewed them bosses every morning about production and work schedules. They had an attitude when you went to work and they would tell you all they wanted to see was assholes and elbows and go get it. It was a rough company to work for and it is a rough, rough business. I was in the wrong place for twenty-eight years, and I know it now. I was the only one in my family that went to the mines.

That was the idea: the big guys would get the boss all riled up and send him down below. Then he would take it

Divided Kingdom

out on the men, and the men would take it out on the work. Their theory was if he is mad, he will work harder.

They would put you in a position to get coal, *regardless of what the mine condition was*. The safety laws said that if you felt a place was unsafe, they couldn't make you work in it. That is one of the first things I learned when I went down in the mine. I worked in the dirt gang, and then they sent me to work for Carl Whitlow as a timberman. I worked for him about three weeks and Benny Wattlet from Stonington was the machine operator. Benny was a little roughneck, too. When we went through the territory, he was out of coal for the day. Whitlow come in there and told him, "You are going to load it out." He says, "If I load it out, by God, you are gonna stand right here beside me."

The first thing I knew, the switch layer come and got old Tony Angraham, who I was working with, to come down there and timber for him. Tony says, "I am busy. I got to finish this rail." He turned around to me, he says, "Take a saw and an axe and an armload of wedges and go down there. They will tell you what to do." I says, "Hell, I never worked by myself timbering. I don't know what to do. I have only been at it for two weeks." He says, "They will tell you. You do it just like he tells you."

This is just what happened. I went down there: Whitlow stood right at the head of the machine operator and they cleaned the slag.* He backed the machine out, and then he would go with me and I would put a row of props* in. We had that place looking like a forest; but, by God, he got the coal out and he stayed there to get it. That's the way he worked the rest of the day. Whitlow was right there with him, and that's the only good thing about it. If you didn't like it and he said, "You stay there," by God, he had to stay right with you. That's why so damn many bosses got their ass covered up, because they would go into places the men refused to go, and they figured they had to have it. They covered a couple

Divided Kingdom

of the bosses up out there running loading machines and the guys would back out. "You've got to get it." Hell, they would take ahold of it and run it and bang, it'd come in on them. They carried a lot of them out of there.

The closest I come to drilling was room neck* and a cross cut*. We knew it was dangerous 'cause we could check it over. We could tell when they were bad, but I went in there. I didn't know it, but the other set of drillers had refused to drill it, too. I told the boss, "I can tell the weight that's on those bars and the way that top sounds if you shoot it you ain't going to load it out." He says, "I got to have it." I says, "Well, I can drill it, but you sure as hell ain't going to load it out." "I will get it, I will get it." I says, "Okay." So me and my buddy, old Henry Hops from Auburn, went in there and we drilled those two places in fifteen minutes and pulled out. The shooters went in and shot the bottoms*, which is below the cut. They shot one snubber* in the middle of each place, at the same time they were shooting two places at the time. When they shot the first top shots above the cut she started popping and cracking, and before long they just went to the last bar in the entry in the cross cut and pulled their equipment back, stood there for about five minutes, and the whole damn works fell in. I mean the whole area, both sections: the cross cut, the entry and the main place in the cross cut there. The whole damn works fell in, twenty feet high. We knew better than to go in there. Hell, we knew it was going to fall in. You could tell. It would talk to you if you listen to it. You could learn to hear that above the machinery. I don't know how a guy can hear that, but you can hear it. If you hear it and stay, then you are a damned fool.

When I got both feet broke at the same time, it wasn't a bad cave in. It was just about a six-inch flake of rock about eight-foot wide and four-foot deep from front to back, but it caught me stooped over. It didn't have to do anything but drop, and we didn't know if it was loose. We checked it and it sounded just as good on the floor as it did on the top. If they are thick enough, they will do that. It just fooled me that time. But now, a lot of those places you can walk in and you can hear them pop if you stand and listen for a minute. If they pop two or three times right quick, it is breaking up pretty bad. Unless you have got some damned good supports in it, you better leave it alone. You learn that quick, or you don't survive.

Paul E. Dixon

Unlike today's mine company, Peabody Coal Company operated solely on a profit and loss basis in the 1930s, which dictated the need to take advantage of the miner's vulnerability and force him to work under adverse conditions. One of the miner's enemies was a pocket of bad air, or "black damp," as it was called. "Stormy" Dixon taught his boss a lesson when the boss insisted on mining where it was unsafe.

I was a door man,* a ventilation man, at first. We shoveled a trench across from one wall to the other wall at the entrance. I had my props set and I got my bottom board on. The black damp* was real bad down there where we shoveled out this trench. I could hardly get down there and get back out in time before I had to take a breath of air. So I would stand there and figure out where I had to drive the nail. I started the nail out there in the good air; then I would jump down and nail the board on. Then I would get back and get another board, go over to the other rib and do the same thing.

After I had it up about four boards high, I stood at the end of the curtain. Superintendent Johnny Hardy was sitting there with the air coming in—the good air—and I stood there for quite a while out of wind. Didn't have any wind at all, so I sat there. He said, "Well, get down there and nail that board." I said, "Man, I am getting a breath of air." He sat there a while and he looked up again and said, "Get down there and nail that board." I said, "I told you, I am getting a breath of air." He said, "If you're not going to do anything, get your bucket and get out of here." I said, "Well, if I am going to go, you better send somebody to guide me to the bottom, because we don't have any light. All the power was pulled and maybe I don't know the way out, and you have to pay me till I get out and get to the top and everything." He wasn't going to send anybody with me to take me out. So he stared at me again. He grabbed my hammer and board and got down there to nail that board on. He didn't even get the nail driven in and he like to have lost his false teeth. He got a whiff of that black damp and he came out of there. He never fussed at me again. See, he thought I was shamming and I wasn't.

In the dirt gang they wanted you to load a car an hour, and if you couldn't hack it they would tell you there were the men on top looking for work, so get in there and get it. All the bosses wanted to see of you were your assholes and elbows.

I was a boss in a mine, too; but I was bossing in the 1930s when it was different than what it is now. At that time the mine had to turn a profit on coal sold; now there is a contract that the utility pays for the costs to produce coal plus a small profit. The difference between them is, if you don't make a profit in the first instance, you don't have a job. With cost-plus they don't care what the cost is because they get ten percent over whatever the cost is. If mining coal costs a thousand dollars, they get a thousand and one hundred. It doesn't bother them, because their contract is with Commonwealth Edison. Commonwealth Edison pays them for the coal; Commonwealth Edison turns around and puts it on the consumer. But now on the profit-loss basis, if you don't produce enough coal and get the sales high enough to where you can pay your upkeep, your taxes, your overhead, your land and your coal rights, then you had to shut down.

The way I used to tell the men that was working and didn't want to do anything, I would say if you had a twelve, fifteen, eighteen-hundred acre farm, and you had three or four men working for you and they didn't produce enough grain, you had to make a profit before you could pay them and pay yourself a little something. You would have to get rid of them four men; that is common sense. You can't live beyond your means. You will go broke. No doubt about it.

Perry Gilpin

As Perry Gilpin relates, in the early 1930s accidents happened everywhere in the mines, but an inordinate number took place in the haulage process—hauling full or empty coal cars to and from the face. Exacerbating the problem were the absence of periodic safety inspections and complete lack of safety clothing.

I quit mining because I got hurt every time I turned around. A clod fell on me once. My uncle drug me out of the mine. That's when I broke my leg in two places— my ankle. Then I was coming out on a motor* and a pole jumped the overhead and hit the ceiling and it come down. I still think it is broke. There's a knot there, you see. And then I got this hand broke. We were out there and a guy hit me with a sledgehammer. I was jacking up the track one night, and the boss hollered at this guy and he jumped and swung, hit the ceiling. It knocked him off balance and the blow broke my hand. I was hurt about four times. I quit. My uncle [who got Perry Gilpin his job] said, "Nothing to it; you may never get hurt again." He went back to work, worked

Figure 51. Underground miners, Peabody Mine No. 58, late 1930s. Notice the hard hats, a new requirement in 1936. (Courtesy of the Boch Brothers)

two or three days and got killed. I don't know exactly how, but I think the roof fell in on him. Had them old-time loading machines, that's how my wife's dad got killed. When they shoot that coal they didn't put enough powder in there, see, and it was too tight. That boom would just fly everywhere, knock the prop out, and the top fell in on him.

The conditions down in Mine No. 58 were rough. They was rough! Buddy, when you crawled off that motor and got to your job, that's when you went to working and you worked until they blowed the whistle for noon. You didn't stand around like you do now. There was damn little safety down in the mines. Only your own safety. You just took care of yourself and everybody was cutting on one another's throat. The bosses are the ones that made the conditions. There's been a lot of laws changed since then. You can ask Max Boch there, he worked for the law, and the only safety

Divided Kingdom

first you got was what you made of it. They had mine inspectors but damn little work they done inspecting. Now, by God, they inspect them, but they didn't then. They used to kill one a week out here at 58. You don't hear that anymore. The government stopped that. They had a wreck out here in 58 on the main line—that was before I went to work—and they had mules. They called everybody out to clean up the wreck and the first goddamn question old Bill Hardy asked was, "Did they hurt the mules?" That's the truth. And somebody said something to him about it later on, and he said, "By God, we have to buy a mule, but you can hire another man." Yes, sir. No, they didn't get much.

Well, we just wore regular pit caps up in West Frankfort early. But the hard hats was in when I went down below here. That was about 1937, 1938. They was in already. Miners liked them. That was one of the first safety measures that I can remember, now since you mentioned it, that ever done the miner any good. Because them cloth caps—hell, I got a scar right there on top of my head when I had a cloth cap on and a clod fell and hit me. But them hard hats was alright. I mean, it protected your head. It's a shame they couldn't have made shoulder pads and all. (laughter) You had to have steel-toed shoes before they'd let you go below in 1933. In fact, if you wore a pair of shoes like this, you'd stumble and fall and mash your toes and get bruises.

Stuart Lidster

Speed was one cause of accidents during haulage*; another was simply the company's neglect of the miner's safety.

The trip rider* rode the front end of the coal car into the mine. He had a whistle and blowed it for the motorman* to pull back a little. When the car was loaded, he blowed the whistle for the motorman to take off, and then the loading machine operator would shut the machine off. You rode the back end of the car out, and you switched it in to cut it off, then back and coupled onto the other loads that you'd already loaded. Did that on a fly if you didn't get squeezed in between two cars. It was timing.

When the car went to the tipple there was an iron lip on the door that when it was hoisted up to the tipple, there was iron things that hung out, and a bar across. The boss would catch this lip and raise that door and the coal would empty out of the car into the hoppers on top.

Well, a lot of times the car would hit too hard and that door would be hanging out. There wouldn't be hardly any bumper left. If you was in there like that, you're going to get

Figure 52. Motorman (driver) on a trip underground in the Moweaqua Coal Company Mine, January, 1933. (Courtesy of *Decatur Herald and Review*)

your shoulders shoved up under your chin and maybe get a collarbone broken. I got squeezed. They got a lot of fingers cut off. That used to be a joke in those days about that a finger's a new Chevy, you know, when Chevys were about five hundred dollars. So if you lost your finger, you got compensated and you could buy yourself a Chevy. That's what some people claim some guys did on purpose, but I never did.

Tom Rosko

In this excerpt Tom Rosko explains why haulage accidents and safety violations were so prevalent—the United Mine Workers did not protect the miners under John L. Lewis.

Well, you're trip riding, you know. I had my feet run over and my hands run over. Not only that, your top would come down. I had two narrow escapes, I'll tell you.

I don't care what they say. There's accidents that could have been prevented, but there was a lot of violations by the company, see. In them days we had no organization in 1925 with the UMW. The bosses done as they pleased. And listen, if you didn't work like they wanted, they didn't care. There was always somebody looking for work, see. So a fellow like me that liked to do a day's work and maybe was more or less afraid for his job, so he'd overdo it.

I had a friend of mine that I was riding trips for at No. 10. I visited him the other day and he's ninety years old, but he thought about it. Well, he forgot that I was riding trips for him. He told me, "Tom, a fellow by the name of Harmon was the boss then." He said he made him so nervous that he quit. Harmon was riding him. Hell, he couldn't get fast enough. So I said, "You're not telling me anything, George. You know I rode trips for you." Boy, he would go like a bat out of hell. And, naturally, he would throw the empties in, or he would say to the guy that would pull them in, "Hurry up, Tom! We've got to go." I would say, "Hell, you just got here." That's how afraid that fellow was.

At that time we only had an organization in name. But we had a fellow here, what they called a mine man committeeman, see. Hell, he didn't have to work. So finally the people got tired of him and elected a good fellow. That's how I got in the labor movement. A fellow by the name of Jim Andrews was elected as a one-man mine committee, see. He got tough with them; things changed when he got on there. He was one Irishman with a temper.

Union Traditions and the United Mine Workers

The miners met the attitudes of Peabody Coal and their working conditions head on. There was a long tradition of dealing with the abuse of coal companies in District 12 of the United Mine Workers. The initial organization was formed during the prolonged coal strikes at Virden and Pana in 1898, struggles that dealt with issues of work abuses.

From its inception District 12 was special. Early district organizers were British trade unionists, sophisticated in political ways, who stressed the class-based nature of union loyalty. The content of their appeals drew upon the tradition of republican idealism and the need for economic equality between the operator and miner. Consequently, the early rank and file were articulate and aware, and believed fervently in the democratic control of union government at both district and sub-district level.

Beginning in the early 1900s, a large number of unskilled peasant immigrants from Italy, Poland, and Russia started to work in the Peabody mines. Their incorporation into the work force was facilitated by the UMW's strong control over wages and conditions of employment. These miners' commitment to the cooperative movement and the union was attractive to the UMW, and their participation in the practical democracy of their local union was especially noticed.

The fusion of these two cultures was mutually reinforcing. The early UMW (1890-1920) embraced both the solidarity of the ethnics and the egalitarianism of older British trade unionists. Both were reinforced at the community level by communal and family sharing and cooperative marketing. As shown in Chapter I, ethnic communities had a long tradition of self-sufficiency dating back to the old country, and habits of mutual support born of a pre-migration economy; all shaped a powerful and persistent sense of community. Beginning in the 1920s, this ethnic culture both reinforced and shared the labor radicalism of British trade unionists—especially an uneasiness with the concentration of economic power that threatened the personal autonomy of the miners. The members of the two cultural traditions felt their communities were threatened by the rapid pace of mechanization and the deskilling of traditional work processes.

Evidence of the power of this new coalition in the 1920s can be seen in the increased pace of work stoppages, community-based protest movements, and attempts to organize alternative unions. The United Mine Workers' rank and file possessed a time-honored tradition of local work stoppages as a means of protecting their lives, unions, and communities. Such actions could be precipitated by the dishonesty of a crooked weighman, unsafe working conditions, an incompetent foreman, the unavailability of adequate tools, or even inadequate supervision. In short, the local strike—or wildcat—was used as a reasonable form of self-protection. In the 1920s, however, work stoppages extended to community-wide issues such as housing rentals, sanitary conditions, and the pricing of supplies.

Most serious were the attempts by a handful of Italian and Polish American trade-union radicals to organize an alternative union in the late 1920s. Radical miners focused their attempts on the Reorganized United Mine Workers, the

National Miner's Union, and the Committee for Miner's Democracy. All failed, but the ethnic miners formed a potent force which kept alive the tradition of rank and file protest.

Considering the degree and extent of rank and file concern over local issues, the union contract was sacrosanct. In particular, it codified the customary pace of work and local customs which protected craft autonomy and in turn assured the ethnic community of its livelihood. The UMW District 12 contract in 1924, for example, incorporated a number of time-honored demands: the weighing of coal by union-sponsored check weighmen; the investigation of accidents; proper ventilation; and specific rules governing timbering, shooting, and loading. Though honored as a tradition in previous years, the 1924 contract formalized the practice of job sharing, or "equal turn." This provision, as pointed out in several interviews, was assurance that the coal companies would not favor more productive miners and discharge those miners who were not. Job-sharing included provisions to share any available work with all the men if layoffs occurred. The attempt of the coal companies to scuttle job sharing provisions in 1924 led to a violent strike which idled the miners for thirty days.

By the late 1920s, union locals' efforts in central Illinois were directed at slowing down the pace of work and impeding the introduction of loading machines. Led by Ray Tombazzi and Joe Ozanic, a number of subdistricts resisted the introduction of wage scales for loading machines in the contract and supported traditional tonnage rates.

Another area of contention which the UMW had formalized into contracts was the grievance procedure. The original agreement established pit committees of miners to represent the union men. Local union efforts in the late 1920s were directed at strengthening the committees and the procedures since the process was corrupted and the committees ineffective. Ineffective pit committees exposed

Divided Kingdom

the miners to arbitrary decisions of management and the erosion of traditional working arrangements.

The concern for traditional work rules and effective pit committees centered on the introduction of coal loading machines that would replace hand loading, the last bastion of the skilled pick miner. With the introduction of the machines, Peabody Coal could, and soon did, divide the work into minute components. The duties of loading would fall on company men who were paid by the hour, as were men who were doing timbering, track laying, and clean up. Miners would lose control of the pace of their work, and how it was to be done.

By the early 1930s, the United Mine Workers proved incapable of dealing with the miners' fears of coal loading machines. Additional grievances centered on seniority rights, effective grievance procedures, grassroots participation, the pace of production, and the corruption of District 12 of the UMW. The state UMW addressed none of these issues. The cumulative sense of frustration and fear over these UMW failures led to a spontaneous explosion which engulfed both the union and Peabody Coal in 1932.

Louis Wattelet

Louis Wattelet—like all Progressives in the Taylorville mining area—recalls the old traditions of the original United Mine Workers and the struggles that preceded the initial organization, especially Virden Day.

That is a big day for the miners—Virden Day. We have a statue out here in the Taylorville cemetery of the coal miner in commemoration of Virden Day. It commemorates the riots of 1898.

I imagine this is why the United Mine Workers first organized. It was all for each other at first, you understand; and then they fought for that. They fought for that right to be unionized. And there was bloodshed at that time, way before my time. Even before my dad's time, because my dad came over in 1900 and this was in 1898 when they first organized the United Mine Workers. They had to struggle, I tell you. It was pitiful—miners in them days. Yes, the United Mine Workers were good for the miners. It is just this damned situation that happened over a contract in 1932. But unionizing the miners was one of the finest things that ever happened. Boy, I will tell you, that was darn hard work for what little pay they got back then.

Ray Tombazzi

In an interview conducted in 1972 by staff at Sangamon State University, Progressive mine leader Tombazzi contrasts the older sense of solidarity of the UMW with the John L. Lewis-led UMW of the 1930s and beyond.

In the early days the UMW union was for the betterment of men's wages and conditions. We were led to believe that if you didn't have someone to relay your grievances, you were at the mercy of the coal operator or whoever your employer was. If you were under a union, you would take up your case with the pit committeeman who was your representative—that's the first plateau—and then if he didn't agree, you could take it up with the president. If the president didn't agree, you could take it up with the board member who was a representative of your district. Then they would go through the group board and if they didn't agree to your problem, they would take it up with the joint group arbitrator. His decision was supposed to be final.

The union was the only recourse we had to get any justice because, as I said before, we were at the mercy of the coal operators. Without a union you would have no grievance committee to take your problem to—no one. Of course, later on, legislation was enacted in Roosevelt's term: the Wagner Act, which gave us more leeway. The National Labor Relations Board stressed the right to organize, which heretofore was a hard thing to do. It was on the statute books, but nobody would give us any support until Roosevelt came in.

The union today is a dues collecting agency. That is its prevalent function. I don't see any unionism at all compared to my days of mining—no, not one iota.

Kenneth Cox

The Kenneth Cox interview disclosed some of the abuses committed by the single man pit committee during the 1920s.

Some of the guys were discontented with the one-man pit committee set-up. We thought a lot of times that a one-man pit committee was too easy to bribe. This is why I felt this way. We had a contract for eight hours of work. When the time came to go to work in the morning, they'd always blow that whistle at least five minutes before starting time and they'd start to wash coal. Us band pickers* had to be there, because coal is coming through the tipple. They was always about five minutes late blowing that whistle for noon and we was always about five minutes late getting out of there to go eat, see. Then they'd always blow that whistle four or five minutes early after the noon hour was over and you had to be there when they started. They was stealing that time off of us, see. Then, in the evening, they'd blow the whistle for quitting time, but they'd hoist coal for about five minutes more and you had

Divided Kingdom

to stay up there and pick band.

They was stealing anywheres from fifteen to twenty minutes a day from us guys. The miners got on Jugglet about it; he was the pit committeeman. Even the men that worked below, they got on him about it and he didn't seem to try to stop it. Now, anytime you're on a pit committee job and the company is violating the contract and you're sitting right there watching them, why should you have to have somebody come to you and make a complaint?

Tom Rosko

Rosko contrasts the difference in conditions under the Progressives and the UMW during the 1930s.

Under the Progressives you wasn't working under a fear. Why, under the United Mine Workers you just were fearful all the time about losing your job. Jobs were hard to get in them days, because that was pretty near the start of the Depression. Rice [owner of one of the Witt-Montgomery County area mines] was a good man. I mean by that he was a good owner and he had good bosses. But their conditions were a little better; you wasn't fearful. If you didn't get the production, they wasn't always riding you to death. And Lord knows that a lot of times people took advantage of it. It's just like everything else, see. Well, he liked the people so well, he used to give us picnics every year.

The reason I know these things [how he fared so well] was because under the Progressives we used to have a lot of what they called seminars. We used to have monthly meetings and exchange ideas. See, like we'd go to Nokomis. They had a certain district and we'd go every month and compare the conditions in your mine and the other mines. Mount Olive did, too. We'd go to the meetings, and then if they had a condition that we didn't have, well, you know, we'd discuss it at the meeting. Then maybe the committee would take it up. That's the way it was. That was all under the Progressives.

We made progress under the Progressives. You take your working conditions. I didn't have to worry about it if I was a little late coming in with a trip or something like that. You didn't have to explain it to a guy or everything else, you see. What I mean by that is, at Coalton the biggest part of the people put in a good day's work. At that time they used to have a limit so more people could job share. They used to have what they called a four-car limit; couldn't load any more than that.

Figure 53. Progressive Miners of America, Taylorville Local officials, March, 1933. (Courtesy of *Decatur Herald and Review*)

Figure 54. Labor Day parade in Springfield, Capitol Avenue, 1935. The line of PMA marchers stretched for over a mile. (Courtesy of Illinois State Historical Library)

Chapter III
The Miner's War:
Defending the Miner's World

I got the rabbit fever out there [in the countryside trying to survive] from skinning rabbit . . . I had to go out in the morning and shoot a rabbit before . . . breakfast. Whatever we could get, that's what we ate. I was doing the Progressive miners.
—**Alvin Wise, 1986**

At the onset of contract negotiations in 1932, Christian County's miners had serious accumulated grievances against Peabody Coal and the UMW and were anxious over their futures. At the same time, they had created a complex world that shaped their loyalties and expectations. Memories of past wildcat strikes, past contract disputes, work pressure underground, corrupt UMW practices, family needs, and communal concerns intermingled in their consciousness, and formed the backdrop of their reactions to the impending crisis.

The actions of the UMW in forcing the new contract on the miners (especially the agreement to use machine loaders), combined with the reopening of the Peabody mines in August, 1932, created an unprecedented reaction on the

Divided Kingdom

part of Christian County miners. In response, the miners organized the Progressive Miners of America September 1, 1932, and conducted a strike against Peabody. Such actions brought the various factions of miners together and raised miners' consciousness to new levels. Behind the actions was concern for the survival of the miners' family and community institutions, which all felt were under attack. The miners had learned early they could depend only on the security of the community, and its security was threatened by the actions of Peabody and the UMW.

Some historians have argued that the initial drive to organize unskilled labor at the grassroots level during the Depression came primarily from the United Mine Workers and John L. Lewis. This is obviously not the case here, since Lewis and the UMW fought the Illinois coal miners on that very issue. Furthermore, our interviews suggest that Lewis and some New Deal Democratic congressmen, such as Frank Fries, the congressman from Macoupin County, Illinois, borrowed ideas from the PMA. Later, they applied lessons learned from observing the strike of 1932 against the UMW in places such as Taylorville to the organization of the Congress of Industrial Organizations.

The major stimulus toward a grassroots labor union was job security. Ray Tombazzi and others placed blame for the foment squarely on the shoulders of "mechanization ... and technology [that] was moving fast." What they meant, as Tombazzi patiently explained, was that "the social economy [enough jobs to support families] wasn't keeping up with the industrial economy." Even more revealing was the analysis furnished by Joe Craggs: "The miners were traditionalists and went by the old ways of the community ... they were threatened by machines."

Of course, the ties of the enclave strengthened the resolve of the PMA. That union drew its strength from the work of the rank and file: they organized the unions' relief

Divided Kingdom

efforts through family participation and sacrifice; they brought unrelenting communal pressure to bear on strikebreakers who were born and raised locally; and they organized grass-roots political pressure on local politicians.

Indeed, the miners who crossed the PMA picket lines felt compelled to explain their actions because of the irresistible pressure placed on them by local strikers. Local miners who went back to work at Peabody mines suffered greatly at the hands of the community. The vast majority of strikebreakers were desperate, out-of-work miners who had been replaced by machines themselves, who had moved from southern Illinois, and who had little idea of the hostility that would be shown toward them. "We didn't come from Sesser . . . to break the union or take their jobs," explained Otto Klein. "We came here mostly because we were starving. We were out of work."

The actual details of the prolonged strike are gruesome. Crimes were committed on both sides. In Taylorville, there was bitterness over the presence of the National Guard, the Sheriff's office, the "goon squads," and the role of Peabody Coal. Everywhere, suspicion centered on individuals who were suspected of committing murder, or who were suspected of supplying information to the UMW. Certain names continually surfaced, such as "Cully" Abrell, "Fats" Cheney, and other hirelings from the southern Illinois coal fields. Sometimes, according to reports in the Decatur *Herald*, the license numbers of cars seen at the site of bombings were traced to members of the old Birger Gang, a prohibition-era gang from southern Illinois.

Aside from these sensational details, the major point is that miners challenged community power structures and the hierarchy of the UMW with their picketing. They had been treated harshly for years, and ignored by the UMW in the 1920s, but did not move against this system until the introduction of new machinery and a labor contract threat-

Divided Kingdom

ened their community base. Miners then were willing to mount a challenge to both Peabody and the UMW—a risk some considered suicidal—and, in the process, risk injury, violence, and starvation.

The strength of the miners' enclaves served them well in their prolonged struggle against Peabody and the UMW; they almost prevailed in gaining recognition at the larger operations. By 1932—at least in the Christian County coal towns—the movement toward open resistance and PMA organization was a strike for economic democracy. It was also a determined attempt to preserve and extend the enclave.

A chronology of the Mine Wars 1932-1934

1932

April 1	United Mine Workers District 12 (Illinois) went on strike.
July 8	Governor Louis Emmerson's Arbitration Commission recommended a basic five dollar per eight-hour pay contract.
July 9	Referendum on reduced five dollar rate rejected by UMW rank and file 4 - 1.
July 13	1,500 marching, striking miners turned back at edge of Taylorville by 200 deputies.
August 10	Basic referendum again presented to the miners. Ballot boxes of all Illinois locals stolen by two armed men in Springfield.

	International President John L. Lewis declared state of emergency and signed the controversial contract, putting the new terms into effect immediately.
August 11	Peabody Coal announced that its four mines in the Taylorville area would resume work on August 12, 1932.
August 18	Striking miners from the Midland Tract and nearby towns marched on Taylorville in an effort to close the mines.
August 19	Peabody closed the four mines.
September 1	The Progressive Miners of America formed in Gillespie.
September 6	Taylorville Local of the PMA held its first meeting in Manners Park, Taylorville.
September 18	Taylorville Daily Breeze and Subdistrict Office No. 5, UMW, bombed. Sheriff Weineke and State's Attorney Gundy requested state militia.
September 19	120 state militia arrived in Taylorville with orders to clear the streets. Three miners jailed.
September 23	Mine No. 9 resumed operation.
October 6	Kincaid High School students held strike to protest use of Peabody coal in furnaces.

Divided Kingdom

October 10   Mine No. 58, Hewittville (Taylorville) resumed operation. Pickets from Staunton, Witt, Pana, Nokomis and Springfield dispersed by tear gas.

October 12   Traditional Miners' Day Rally to commemorate the martyrs of the Virden Strike of 1898 held in Taylorville. Three thousand protestors engaged in disturbances with state militia.

October 13   State militia shot Andrew Gyenes, who was a PMA miner in Tovey and a mail carrier in Langleyville.

October 16   Andrew Gyenes funeral in Manners Park.

October 25   Peabody Mine No. 7 at Kincaid resumed operation.

November 1   PMA Women's Auxiliary and other charities distributed provisions to striking miners' families. PMA soliciting committees collected food from area farmers.

December 17   Chicago and Illinois Midland Railroad bridge destroyed by explosives.

December 23   Last contingent of state militia withdrawn from Taylorville.

1933

January 3   Protracted gun battles between 150 Progressives and 50 mine guards at Peabody Mine No. 7 at Kincaid. Emma Cummerlato

	and Vincent Rodems were killed, 12 wounded, and 30 arrested.
January 4	James Hickman, mine boss at No. 7, killed in a gun battle.
January 5	Five companies of state militia recalled to Taylorville.
January 8	Funeral of Emma Cummerlato.
January 11	Circuit Court restrained Sheriff Weineke from interfering with operations of PMA relief stations in Kincaid, Taylorville, and Langeyville and with regular PMA board meetings.
January 12	Illinois' new governor, Henry Horner, met with leaders of UMW and PMA. They agreed not to picket in Christian County and that the National Guard would be in full charge of law and order.
January 26	Nearly 10,000 miners' wives and miners marched to Springfield to protest practices of UMW.
March 10	Mrs. Andrew Gyenes sued the state and Sheriff Weineke for $10,000 for the shooting of her husband.
April 6	Miners involved in gun battles of January 3 and 4 acquitted for murder of James Hickman.
April 19	Circuit Judge Jesse Brown issued temporary

Divided Kingdom

	injunction restraining the sheriffs of five counties, including Christian, from interfering in the activities of PMA.
April 27	Two bombs partially destroyed the home and demolished the garage of John Stanley, president of the Midland local of the PMA. This is the thirtieth bombing in Christian County.
May 18	In Saline County, PMA miners refused to work at Peabody Mine No. 43 and 47.
June 12	Byron Haines, face boss at Mine No. 7, attacked and badly beaten by six Christian County men in Mt. Olive.
June 30	PMA meeting suspended in Taylorville and Tovey by county officials.
July 12	George Mosey, Manager of Peabody No. 7, and Bill Daykin, PMA leader, fought in the square in Taylorville.
July 14	The National Recovery Act Code for Coal established eight-hour day, basic wage of five dollars per day. The PMA was excluded from negotiations.
July 23	Andy Newman's, Jack Stanley's and Leal Reese's houses were bombed. Car used in bombing traced to "Cully" Abrell, a gunman hired by Peabody, and "Fats" Cheney, a bottom boss at Mine No. 58.
July 27	Bus of state militia pulling out of Taylorville bombed. Two dead, twenty injured.

Divided Kingdom

July 29 — John Wittka, Sr., and Ben Denoff, Langleyville, beaten by a gang of PMA miners. The two had just begun work at No. 9 the week before.

September 20 — In a major address to 10,000 unemployed in Christian County, Agnes Burns Wieck, president of the auxiliary, PMA, denounced the eight hour day and promised "turbulence" in the coal fields for years to come.

October 2 — Taylorville PMA petitioned for their old jobs in Peabody mines. Petition denied unless miners affiliated with the UMW.

October 6 — Governor Horner ordered closing of Mine No. 43 in Harrisburg to avoid bloodshed.

October 12 — John Stanley's house again bombed and his bodyguard shot. Indictments are returned against "Fats" Cheney and "Cully" Abrell.

December 23 — Gun battle outside the Brass Rail Beer Parlor, a PMA saloon. Dominick Hunt, a PMA miner, died December 28.
Jack Glasgow, Pat Kain Jr., and Tony Roazenski wounded.
Glasgow charged with murder.

1934

January 2 — Virgil Tamburini's residence in Langleyville bombed. Bomb may have been intended for Ben Wenoff.

Divided Kingdom

January 5	Coroner's jury returned verdict charging Glasgow formally in the death of Hunt.
January 10	In Chicago U.S. Court upheld the UMW contested contracts with the coal operators.
January 18	PMA commissary at Kincaid bombed. Fifty-fifth reported bombing.
January 19	In Washington, NRA ruled in favor of UMW wage contracts in Illinois.
March 27	National Guard withdrawn from Christian County.
April 26	Parade celebrating the election of three UMW members to Kincaid board ended in violence. Sam Roncheti, PMA, and Frank Angenednt, UMW, died.
June 18	In Taylorville, the Glasgow murder trial opened.
June 26	Glasgow acquitted.
November 10	Ed Manvel defeated Betterton, UMW candidate for Mayor of Taylorville, Manvel pledged to bring law and order to town.

Causes

Harry McDonald

Changes in consumption, mechanization, and periodic unemployment were the underlying causes that led to the changes instituted by John L. Lewis. Harry McDonald suggests that the strength of the PMA resistance came from the most hide-bound, traditional immigrant miners who did not wish to see changes in coal mining.

The improvements in the coal mines in those days were probably what ignited all the union troubles. John L. Lewis was the president and he believed in progress. I always looked at it that way. He wasn't trying to hurt his people that he had organized. He was trying to show them a better way. To progress, you had to accept it and the coal industry. The operators were bringing in the people who had ideas, inventing new and better ways to mine this coal.

Illinois was where changes in coal mining occurred. Bigger equipment, automatic equipment, or whatever they were improvising, was focused here. One of the main things was the loading equipment—different types of loading equipment that was improved on—and today has no resem-

Divided Kingdom

Figure 55. Peabody Mine No. 5 in Pawnee was abandoned in 1908, and dismantled in 1921, a victim of inefficiency and lack of demand. (Courtesy of the Boch Brothers)

blance to what they started out with. It was something that was necessary for the coal industry to survive. They had to be able to get better production. If we'd never changed the old plan that they had, they'd never be able to produce the coal that is needed today for our power companies.

Like I said before, the early coal demand was mostly by families—home heating, and business building heating—and it was governed by temperatures outside. Just as soon as the temperature got warm, the people just didn't buy coal any more. These coal mines then shut down, and there was always an extra cost for the operator. Well, those shutdowns were eliminated with the coming of the need for electric power—the automatic equipment and better equipment to work with—which I still agree was the best thing. The mines worked continuously supplying coal to power companies with automated machinery. The miners that wanted to stick with their old methods were very radical. They created trouble all the time for people in the small towns who didn't want to go along with their resistance.

The people who were for the improvement plan suffered from changes, because they were in a minority, really. It was their ability to accept changes that put them apart. Maybe the second generation would be for improvement, where the man who migrated here still stuck to the old style that he knew. He didn't know the new, so he stuck to the old.

There were many people who migrated here in the early part of the century just for that reason. They made their living here, and that was their history. [Coal companies] went overseas and solicited them to come to the United States, either to be union strikebreakers or just be employees; they used them any way they could. Do you understand what I mean? If we had trouble here, they would bring these people in because they could not speak English, and because that was work and they took it.

The people most apt to resent the modernization move would be the first generation era, depending upon their age. Some of the older ones had sons that would still resent the move to modernize and they would be in sympathy with these old people. It was definitely first and partially second generation people who disagreed with the change. See, the older ones accepted the old conditions; they knew how to do that work and they didn't have an education. You'll notice that your first generation to come over here didn't educate their children like they do today. It was an important issue. The coal-mining families had to resolve [this lack of education] because of the changes in coal mining. It wasn't only in the coal industry; it's in everything that we do that there's been improvements and advances in our way of life.

Divided Kingdom

There were five coal mines in this Pana area. Most of those people—or I'd say all of them—went to this rival union that did not want improvement [the PMA]. All these mines were old hand-loading types, and they were operated by hometown owners. But they were small-time people and they kept the coal mine in this area on that low scale.

All these people went Progressive in this area [Pana]. The only ones who were United Mine Workers might've left here and gone to another mine in another town, but there weren't any United Mine Worker mines here after the Progressives came in. There might've been before the Progressives, but in that change, they all went Progressive.

These people here, their wives and the husbands, would come up to Springfield in the 1930s and they'd march along with Stonington miners, the Virden area people, and all those small towns that had Progressive backing. They were strong, and the United Mine Workers just couldn't operate coal mines in that area. Peabody Coal Company did manage to operate on the Midland Tract, around Tovey and Bulpit, and Langleyville. They managed to operate those mines under United Mine Workers. But that's where there was turmoil all the time; even families separated because of their commitment to one side or the other. You know, one would work and the other didn't believe in working that way. They'd be mad at each other because of jealousy or hatred or whatever. Brothers fought with each other, that's just how bad it was.

During this time in the 1920s when this Progressive movement was on the way, the Peabody Coal Company owned a coal mine in Riverton, Illinois. The town area was settled by immigrants from Italy and probably some French. They had a huge company area with houses where the people lived and worked for the coal mine. The union wars shut it down and their production was hurt so bad that Peabody Coal moved their operation to other towns where they could accept the machinery. I don't think they ever did get any machinery in that mine at Riverton.

Another problem was that the Progressive mines folded. The Gillespie area and the Carlinville area was strictly a Progressive area, along with Pana, Virden and all those places south of Springfield were all Progressive strong. They suffered a lot from it because of the shutdowns. At Auburn, the big coal mine was for the Progressives up to their neck—the Progressive movement—and the mine never operated anymore since they weren't profitable. It was the loss of coal mines that really hurt the Progressives.

The United Mine Workers were an international concern, not just a local thing like Springfield or like Illinois. Their leaders [PMA] weren't qualified to do the job necessary to make it profitable for the coal companies and the people working. If they weren't going to go along with the progress movement of better equipment, well then, all you're going to have is low production. So what it winds up is, it couldn't have been a profitable thing for the Progressives to win. United Mine Workers won back all the rights and got their mines going again. We couldn't live today under the production of the Progressives, because we couldn't do it by hand. It's the same way in any industry if you don't have automation. I think that automation doesn't necessarily knock everybody out of work; it provides jobs other places. If they have to re-educate to get into a different type of work, they have to do it, that's all. You can't go on under some old beat-up way of life and get anyplace; you have to go with progress.

Divided Kingdom

Sam Taylor

Sam Taylor's explanation of the precipitating causes of the Progressive Miners' split with the UMW is the most generally circulated story in the community.

You see, what John L. was doing in 1932 was the same thing that the United Auto Workers and the Steel Workers are doing the last two or three years. That is, they were accepting a cut in their pay. Up until that time, all of the United Mine Workers contracts were voted on by the rank and file, and they always would get a little raise. John L. and his buddies worked out a deal with the mine owners. Well, this particular time [1932] the miners didn't get a raise, and John agreed to cut their pay a dollar an hour. The miners were not educated that cutting their pay saved the coal mine owners. For one thing, F.S. Peabody had a private railroad car, and if you were a coal miner, would you particularly be pleased if they cut your pay a dollar an hour so that F.S. Peabody could run around the countryside and look at his coal mines out of a private railroad car?

Anyway, the 1932 contract was voted down all over Illinois. So he went back and got some fringe benefits, but still had that dollar off. They voted on it again; that's when the ballots were assembled in Springfield in July, 1932, and somebody broke into the mine office and stole the ballots. It was John L. that then declared an emergency; he declared the contract was okay and signed it.

That's when the fire [protest] broke loose in Macoupin County and spread to Christian and a number of other counties.

Joe Craggs

Craggs relates another part of the folklore miners created to justify their break with the UMW. The Mulkeytown caravan massacre and John L. Lewis stealing the ballot box of the August 1932 election were staples that strengthened the resolve of striking miners.

In the 1930s, there were two competing unions. The National Miners Union was the first big strike here in Christian County, which came in for about six months in 1930. That union disappeared. Then your next strike was the Progressive Mine Workers of America, which originated in Benld. They were not so easily quelled. That was what we called the big miner's strike and it was real rough; they brought in the troops.

During the National Miners strike they [the striking miners] would bomb our home and they would also shoot any cars that went toward the coal mine. They put bullet holes in them. I think the guy is still alive [one of those shot up]; he used to work for the National Mine Service and he would count the bullet holes in his car. He didn't get wounded. He had 342 holes in it one day going to work.

The Progressives tried to picket and shut down all the mines. They had picket lines and, I think that the biggest thing that broke up the Progressives in this county was that they all decided to go to southern Illinois in 1932 and picket the mines. So, they loaded trucks and cars and their chuck wagon. They started out and they got down just below DuQuoin, Illinois to a little town called Mulkeytown. In Mulkeytown the sheriff and his deputies from Williamson County were waiting. They turned over a truck, they shot them, they beat them, they chased them all the way back to central Illinois at that time, and that did more, I think, to break up the whole thing in this part of the coun-

Divided Kingdom

try than anything else.

Most of them run through cornfields back to central Illinois. You get some of these old Italians out here to tell you about it; they were in it. There is a guy; I think he is dead now—Joe Verardi. I used to kid him about being in the Mulkeytown riot because he came to work after that. That broke up a lot of their Progressive feelings and a lot of them came back to the United Mine Workers fold.

The cause of much of the trouble were old superintendents. The older superintendents had been around for years, and they had kept their cliques going. They were the superintendents back in the thirties and before that they were God Almighty at the coal mine. They were the people that hired and fired-and I mean fired.

There was some blacklisting naturally. There were files kept on troublemakers and so forth; there had to be. During

Figure 56. Caravan of Taylorville and Pana miners on their way to Mulkeytown, August, 1932. (Courtesy of *Decatur Herald and Review*)

Divided Kingdom

the mine wars these coal mines in the midlands were probably the most productive coal mines in the United States at tons per man, and had less trouble with production, and, therefore, they worked steadier. They worked steadier, and the coal miners in this area had more pride in their own self than a lot of areas that I worked in. Also, they had pride in their homes and everything—in keeping them up and things like that. So, consequently, you weeded out the deadbeats and the troublemakers. If that is blacklisting, I don't know. What it was was good management.

Ray Tombazzi

Ray Tom, as he was called, had the keenest insight into the UMW/PMA conflict. He cites the details of the Mulkeytown "Massacre," the problem of the National Guard, the pitched battles at Peabody Mine No. 7, and the thugs hired by Peabody Coal, as reasons for prolonging the conflict. Based upon other interviews and court indictment papers, this explanation seems to be the correct one. His insight that the "social economy wasn't keeping up with the industrial economy" was the best single summary of the underlying causes of the conflict.

The PMA were strong in Taylorville but the element from the outside [strikebreakers] came in and took our jobs. Then, gradually, the home towners were inveigled into going back to work. Of course, I tried to settle in 1936, but there was no way to settle; we weren't making any headway in reaching an accommodation with the UMW. We were at a standstill. We would have whipped them into submission if there hadn't been any collusion between Lewis and the Peabody Coal Company. The United Mine Workers throwed so much money in there and it turned about-face against us—it was all lost. They [Peabody Coal]

181

Divided Kingdom

Figure 57. A road into Taylorville blocked by striking miners, August 8, 1932. Peabody's operations ground to a halt in early August, when miners could not pass the picket lines. (Courtesy of *Decatur Herald and Review*)

Figure 58. The streets of Taylorville were not safe for striking miners. On Virden Day, October 12, 1932, National Guard troops herd strikers away from a meeting in Manners Park, Taylorville.

was on the verge of signing with us before then.

Lord knows how much money was throwed in. They had so many men on the payroll. The coal company wasn't producing any coal at all and, of course, there was a lot of sabotage going on, costing the coal company all kinds of money. The railroads were being stopped at a lot of places. We had the militia from 1930 to about 1934. You might call it four years of military rule. We had about two weeks of martial law there when they arrested so many.

Strikebreakers for the Christian County strike were recruited from West Virginia, Kentucky, and Alabama. They solicited from all over; they had agents going out and recruiting men to come to work, not knowing what they were getting into. They gave them a cock-and-bull story—it wasn't what they thought it was at all. I talked to a lot of them and they said, "They didn't tell us there was a war here or anything." They kept it out of the papers. There was some excerpt or something on the back page of local papers down south whereas it was on the front page of the St. Louis papers. The Chicago papers didn't play it up too much, but the St. Louis *Post Dispatch* did and all the local St. Louis papers. They played it up pretty big. There was about twenty-some miners that was murdered, and I don't know how many homes were bombed. The UMW had PMA attorney's [homes] bombed and a lot of them shot; there were riots in the streets and all kinds of pistol battles. I got shot once here—got a bullet in my leg—and I got bombed twice.

Mulkeytown was a turning point that made the strike more militant. Too, the militancy occurred when these infractions were imposed on us, like curfews, and restraining us from walking in the streets. They had the militia with bayonets and they'd stick you with them damn bayonets. They would arrest people and put them in the courthouse in Taylorville under military arrest. They would load them up on cars and take them twenty or thirty miles out. Then we'd send a truck and bring them back in. This went on continuously for two days.

When local officials in 1932 jailed PMA miners, they [PMA strikers] depreciated the Christian County courthouse by about a half-million dollars—tore out windows and all the library books, and all the furniture in there, and they throwed it out the window. The judge's chambers was all torn to hell. They just demolished the whole courthouse. They would arrest them and put them in the courthouse; they didn't even have jail cells for them. They arrested everybody. I remember they arrested the mayor, whose name was Armstrong, and he was so mad that he said, "Tear the son-of-a-bitch [courthouse] up!" Oh, I tell you he was mad. Of course, later on, Argus, a superintendent at Peabody Coal, got Spresser in there as mayor, and from then on it was sixteen years of Mayor Spresser. He [Argus] had the whole sheriff's force; he had everybody. Everybody was under this thumb, all on the payroll. Finally they broke the PMA up.

The National Guard helped encourage the chaos by their orders. They were ordered to make us move on with no loitering on the street. They controlled the streets all hours of the night and day and picked up anybody that was suspicious. They'd take them up to the tribunal—the military court—where they didn't do anything but harass you.

They took me up there every time they would see me. They accused me of passing ammunition, and the commander would say that they were going to hold me responsible if events happened over there and all that malarkey. Of course, they had goon squads going out—the United Mine Workers—and they'd shoot into homes. If anybody came out alone, they'd beat the hell out of them. Of course, we had our squads and we'd shoot back at them; we'd waylay them. We killed a few of them, you're damn right. We didn't fool around!

We had a battle over there at No. 7 one time, and Lord knows that the ambulances came from all over. Also, we knew that there were three of them that bombed Stanley's, the local PMA president's house, a fellow hired by the UMW by the name of Cheney, who was supposed to have gotten shot; a fellow by the name of Mitchelson; and another fellow by the name of Chuck [Abrell]. They shot his leg off, but he's still living. The other two died. The FBI was supposed to be looking for them for four years and they never did find them. They were hiding from one joint to another and they never did catch them. They were hired—known gunmen—and they were kept in seclusion by the Peabody Coal Company. Their wives said time and again that they were getting paid by the Peabody Coal Company while their husbands were gone.

Figure 59. Striking miners incarcerated in the Christian County courthouse, October 16, 1932.

Divided Kingdom

Figure 60. Striking miners await arraignment in the Christian County Courthouse, 1932. The Circuit Court indicted hundreds of miners, then released nearly all. (Courtesy of *Decatur Herald and Review*)

Figure 61. Strikers at Peabody Mine No. 7, Langleyville, block workers from entering the mine, August 22, 1932. In early January, 1933, the PMA and UMW fought a pitched battle with machine guns at No. 7. (Courtesy of *Decatur Herald and Review*)

The Progressive Miners

Cuthbert Lambert

Many miners were forced to make a decision when the strike began: keep working in the mines, as a UMW miner, or quit altogether. The pressure to stay away from the mines was great, as Cuthbert Lambert relates in this story about his family and the other ethnic miners who joined the PMA in Taylorville.

When the mines went on strike in 1932, mother didn't want him [dad] to work in the mine anymore. He was going to go back, I remember, in 1935, and she politely told him, "If you go back to coal mining, you get somebody to pack your bucket and wash your pit clothes. I won't." So that took care of that. He done odd jobs, and we made a living. We never had any relief from the government or anything like that at any time, and we made a living. There were five kids and mother and dad. He maintained his property—not the best in the world—but we finally got it painted and straightened up and he done a little remodeling. He eventually went to work at the paper mill, and when my older brother got out of school in 1934 he started at the paper mill, too. Then when I got out in 1936 I started at the paper mill, so we made it through the 1930s pretty well.

Divided Kingdom

He left the mines because of the strike. They had a split decision. It was my understanding that the Illinois miners voted John L. Lewis out, and when they got ready to recount the ballots they couldn't find them. Then they [the UMW] ordered the men to go back to work that summer, late, and they had guards out there. The men—at least my dad and his family—didn't want to work with gun protection. They said they didn't feel like it was needed. But none of them went back to the mine.

I remember one time standing next to my dad when UMW representatives Roy Starks and "Fats" Cheney come to the house to talk to Dad to get him to go back to work. He said, "If I come out there, hell, I might not get home. How am I going to get back and forth to work?" "Oh, well, we will take care of you and see that you get back and forth to work. If there is such a thing that you can't come out, we will see that the family is protected," they said. Dad stood and told him, "No, that just doesn't sound like what I want to do. I don't think I am going to get involved in it. I don't need a job that bad. We are making a living." We were, so that was dad's attitude toward them and they come twice; that was all they ever come and ask him. But they had a lot of people here that were rough.

My father was as active as most of them in the Progressives. They went on picket lines and things like that. He wasn't involved in some of the things that some of them were supposed to have done. He stayed home and took care of the family as much as he could. He started working people's yards and trimming hedges and then cutting grass and things like that. He made enough; we got a living. See, my mom was upset because of all the turmoil they had, and none of the rest of them were going back except some of the Wilson boys who were related to us.

The whole family belonged to the Progressive union. They all went to the meetings, went to picket lines and things like that, but I don't think they ever participated in violence.

Now, one of our family—Joe Wilson and his brother—got into a lot of fights. Joe was back over there in Mine No. 58 visiting with somebody and started home one night. About fifteen, sixteen UMW guys with pick handles got him and damned near killed him; he turned black all over. They nearly killed him. After about three years, he did die.

Joe took it upon himself to be a one-man army. He whipped about half of the United Mine Worker officials around here at one time or another. He could do it; he was a hundred-and-ninety-five pound wildcat that nobody whipped. He just cruised around town and when he'd find one of them scabs by himself, he just stomped the hell out of him. And then he would have to turn home and he would have to crawl down the ditch along the railroad track to keep from getting shot by UMW thugs. Because they figured when they got him out the edge of town, they would shoot him. He crawled down that ditch till he got in front of the house, and he would run like hell and get in the house. But he whipped a lot of them, I will say; he got his share.

Most of our neighbors were Progressives. Very few people went to work after that [the strike] and ultimately most of them moved away. My dad was really about the only coal miner that stayed in the neighborhood after the strike.

Up on Ash and Snodgrass, that's where the action really was, because that's where Jack Stanley, the president of the local Progressives, lived. I don't know whether anybody has explained to you about him or not. Jack Stanley's house was blown up two or three times. Twice, I know they tried to get him. I think the first time they blowed the back porch and figured he would go out the front porch and they would get him out there. They brought a lot of them thugs in and it really wasn't safe.

They only got after me once. I don't know if they knew who I was or not, but I left town one night about nine o'clock and I noticed every time I would turn a corner there would be two guys at the corner opposite, turning my way. In other words, they were tailing me a block behind, and I got out at Park Street by a place where old man Whetsall lived. I turned north there and got to his barn, and that was a dead end alley. When I got down there in the darkness of that barn I run like a young rabbit, hit that vacant lot down there at the end of that alley, cut across to Snodgrass Street and went home. But that is the only time I ever knew they were going to start working on us. They had just shot a girl down in Mount Olive that was doing her homework at the dining room table.

My family had qualms about what was going to happen to the kids. Whenever that shooting started up at Jack Stanley's corner, why we had a center closet in the house and dad would always get the shotgun and lay on the floor there at

Figure 62. Decatur National Guard soldiers quartered in the Jury Room of the Christian County Courthouse, December, 1929. The unit occupied the same quarters again in 1932. (Courtesy of *Decatur Herald and Review*)

the back door in case anybody come down that way to our place. We would all go in the center closet so there would be at least two walls, maybe three, to get through to get to us. They shot some pretty high-powered stuff around here.

They had the militia here, but it didn't do any good because they had lists of people and cars that weren't to be bothered that they passed out to each militia group which came in. The only militia group that wouldn't play games with them was the group out of Chicago, and they were only here two weeks; they never did come back. The militia come in here and were told what to do and when and who to do it with. The Chicago group said, "No. We were brought in here to keep the peace and we are gonna patrol the streets and there ain't nobody tell us a damn thing. We are going to do it our way." See, Peabody officials—W.C. Argust, the oldest Start boys, Roy and Lloyd—all of them were involved in telling the militia what to do. It got pretty rough for awhile.

No one could protect us. There was a lot of rumors that the sheriff and state's attorney got as much money from the coal company as they did from the county. At that time they did do some gambling in the county, and the state's attorney got a cut. Then, a few years later, one of the sheriffs said that if he could be elected one term, he would have enough to suit him. One term was all he got; his boys were both pretty good lawyers here in town with a lot of money. That's the way the cookie crumbled.

Ultimately, this Jack Stanley went to the pen for four years for blowing up a bridge over at Pawnee—him and three other guys. They had a short kangaroo court up here for them, and they went straight to the pen for four years. Then, while I was working at the mine, a guy told me, "Hell, them guys never blowed that bridge up." He said, "I know who did it!" He never did say and I didn't ask him [laughs], but I knew they [the Progressives] didn't do it.

Strikebreakers come in here [Taylorville] by the droves

Divided Kingdom

and any four, five rooming houses would have anywhere from one to a half-dozen strikebreaking families living in it. Some of them would work the clock at the mine and rotate shifts and they lived a pretty precarious, rough life. I think most of them come in from southern Illinois and Kentucky, and they were pretty rough people to deal with, 'cause they were up here to get out of what they called a woodsman's ghetto. They didn't have anything to lose, so they come here and had everything to gain when they went to work at the mines. That's why they could buy guns and everything out at the mine, because if they didn't have what they needed, why, they could buy their ammunition and take it off their paychecks.

They had a place there on the corner in Taylorville that an old lady run; I believe her name was Raney. She had a bunch of little shacks behind there and at one time she had them wall-to-wall living in the basement and in the two upstairs rooms of the tavern. Then some of them lived in what they called the chicken houses 'cause they were just little two-room shacks.

There was a dividing line with the strikebreakers in Taylorville, and they kept to themselves pretty well. Of course, they come in here for that reason, and the people uptown [businessmen] treated them alright 'cause they were the only ones working—making money, naturally. All the businessmen went where the money was, naturally. During the Depression, people like my family didn't associate with them if we could avoid it. I did eventually, and my mother really ripped me up the back 'cause I went with the Chief of Police's daughter in 1938 and my family didn't like that a bit.

Well, there was cliques during the mine war at Peabody. The Starks and their friends and the guys that worked with them got favored by the UMW; even after the strike, you could tell it. Guys like me, we had to work twice as hard for what we got. There were things out there I wanted, and they would never give me anything but a drill until they got the continuous miners in there.

During the strike the Progressives set up a relief of their own. They did get some government pork and some surplus dried beans and things like that. Some of the men got on WPA and CWA; but, funny, my dad couldn't get on any of it. The Red Cross did give some help, but I can remember back when the men come to ask my dad to go back to work right after that, some women come out and said they were from the Red Cross. There I was with my dad out in the front yard cutting grass or doing something. Anyhow, we talked to those ladies and they said, "Mr. Lambert, do you own that big house?" And he says, "I sure do." She said, "Well, if you own that big house, we don't need to help you." He says, "Well, you don't need to help me," but he says, "I sure as hell can't feed that house to my kids. We need it to live in, and it would be a little bit hard to digest." That was what he said; that was as close as we come to getting any help. I don't think there was any church organizations going at that time that helped anybody. If they did, we didn't know about it. The PMA Women's Auxiliary did get some government surplus somehow or another, and I remember they give them some pork, too, but you couldn't eat it if you boiled it for a week. It was packed in salt and stored, and it was so damn salty. They burned it; they had to; they couldn't eat it. They couldn't get the salt out of it. So, consequently, they didn't get very much food.

Now, at one time, they had what they called the Civilian Conservation Corps. The young men went to camp somewhere in Wisconsin or northern Illinois—wherever they were doing forestry work. They got thirty dollars a month, something like that. I don't know what the pay was, but they could make a living and send a little bit of money home to their parents. That helped some, but none in my family got into that except, I think, Joe Wilson. That was basically the way things were. I think the Salvation Army did do some good for a lot of people, but the Red Cross didn't help everybody. It was for some people and not everybody.

Most of the north end was Progressive coal miners, I think, and some of them did go back to the mine. But the ones that come in here to work during the strike didn't stick around the north end places like that too much. They mostly congregated close around uptown in the older houses that were rentable. Miners had moved out into different neighborhoods and rented the old places like that. They—both the UMW and Progressives—most generally stuck together, because they had to. There had to be some protection someway or another 'cause there were a lot of people carrying guns back in those days.

I think they [the Progressives] kind of looked after one another. Of course, the old folks that were here, like my family, had their own homes; they would just sit around in their own neighborhoods. They were kind of clannish as coal miners, anyway, I think, and they basically stuck together all the way, until some of them had to leave home to get jobs. When they got jobs—just like my dad's sister, Aunt Molly—they moved to Pekin because her husband had a job working in a foundry. He come back and forth for awhile until he decided, "Hell, that wasn't the way to live." So he just moved up there. A lot of them went to Detroit and got jobs in the automobile factories and places like that. A lot of them, Progressives—I don't know where they went. They moved 'cause the kids that were in school the same age as me basically left when they graduated from high school.

Most of the Progressives were Italians, and there were some Irishmen, a lot of French people and a lot of English people. I had French people and English people in my family, and they were, basically, I would say, the coal miners that were in the community. The majority of them were of European nationality. Italian, Sicilian, English and French.

Divided Kingdom

Progressives were probably a little bit of everything.

I would think the ones that really hung into the union—the UMW of A—and done a lot of the dirty work were basically the people that their families had been here for a long time and were pretty well Americanized.

Ada Miller

Communal cooperation and solidarity were hallmarks of the Progressive Miners' movement. PMA communities with resources sent material and money to PMA mining communities on strike, as related by Ada Miller, the wife of a PMA sympathizer.

The miners formed a new union which was called the Progressive Miners of America, and their intentions were to get rid of John L. Lewis. Since I was in the grocery store and we had them as customers, why, you felt like you had to go along with the rest of them. I joined the Auxiliary. The women had their auxiliary and they had their first meeting in Springfield at the K.C. Hall. I was one of the delegates to go to that in the fall of 1932. We would go and load up on the truck and go on these picket lines, you know; I would do that, too. Get up and be there by five o'clock, trying to stop the [UMW] miners from going to work.

If you didn't get involved, they just thought maybe you didn't care. To show them that I did care, I got involved. There was no reason for me to get involved; I didn't have anybody in our immediate family that was a miner.

We had quite an infiltration of people from southern Illinois that came over and took their [striking miners'] jobs. The strikebreakers were also called scabs and swampies. The people that came from southern Illinois were called swampies. Most of them lived in Taylorville, but not in Langleyville. All the people in Langleyville owned their own homes. No, they didn't come to Langleyville.

I say to this day, I think I am still a Progressive. I think that it was a good thing that they started, but I guess that they just weren't strong enough. The strike just went on for weeks and months and years, and it finally broke their backs, I guess. And we were under militia law then in 1932-33 for about eighteen months. I remember sitting in front of my store in Langleyville, and there were these militiamen on horseback, who ordered me inside. I don't know why, I just did it (laughs). He was on horseback and he just said, "You go inside where you belong." And that was just what I did. But I wasn't hurting anybody, or doing anything. I was just sitting out on the steps in front of our store. The militia were for Peabody, you know.

There was an awful lot of bombings of homes that happened, especially up in the north end of Taylorville. Some of the fellows [Progressives] were Peabody stooges, too, you know. They would bomb their own homes to show that they were on their [Progressive] side. And all the time, they were just Peabody stooges. You found out later on, after it was all over with and it was settled. This one fellow in particular, he just went along with them, and would go to all the PMA meetings and things and just be real aggressive; he would get up and make speeches and all of that. He was Peabody man all the time, and they found it out later.

Well, Raymond Tombazzi, for one, and Joe Gherardini in Langleyville were leaders. Well, Agnes Wiecks was our first Women's Auxilliary president. But we had several. A lady in Kincaid [Ruth Besson] would get up and make speeches. And a lady from Stonington would get up and make speeches at these rallies.

We would hold dances. The women would bake pies and cakes and different things, you know, to sell and make a little bit of money. But, like, if the children needed a pair of shoes or something, you would give your order in and they would buy shoes for the school children.

A bunch of people would get on these trucks and then we would go to different mines and just stand there. Of course, we didn't stop anybody from going to work, because they could go to work just the same as we did. We didn't stop any of the coal miners, I don't think. We were just trying to shame them or something. You know how picket lines are. A lot of people won't cross a picket line. A lot of good union men just will not cross a picket line.

Ray Tombazzi

The PMA established commissaries in the strike districts, which, in turn, distributed food, wood, and clothing to destitute PMA miners. Ray Tom helped distribute such staples to Staunton-area residents.

The PMA had a commissary in the Taylorville area. The other local unions that were working would contribute so much of their pay. I believe the Springfield local here used to give five dollars a member a month. Down in other areas, they would give one dollar a pay or two dollars a pay. I just can't enumerate the amounts. These would be portioned out to various commissaries, and it was prorated according to families. For instance, a person who had five in the family would get probably three pounds of beans, where a person with two in the family would get one pound of beans. Everything would be proportionate according to the size of the family. Of course, we used to go out and ask farmers if they wanted their corn picked. We would pick for a portion of their corn; then we'd take it and get it ground and we'd have corn meal.

No one really starved during this time. Of course, there

were more mealtimes than there were meals, in some cases. If they needed some footwear or something, in my area, around Staunton, they would come to me or put it before the commissary board. We'd make arrangements to give them something to relieve the agony. We would prorate wood. Everybody that worked got their wood; they'd get the proportion of their wood hauled to them. Trucks would be donated and everybody would help in sawing and cutting. We'd clear off places if a farmer wanted some land cleared off and we'd take the wood.

Frank Borgognoni

Langleyville, Bulpitt, Kincaid, and Tovey were Progressive Miner strongholds, and dangerous places to be a UMW miner in 1932-35. Frank Borgognoni's father owned a bar in Kincaid where the PMA congregated.

Well, the biggest percentage of miners in Langleyville were Progressives. Just a few went to UMW. The biggest part of UMW miners had to move out of town. They had to move to Taylorville, or somewhere like that, where they wouldn't be bothered a lot. In Taylorville there was more or less the strikebreakers up there, because they had good law enforcement there. That was the sheriff's office, you know. Sheriff Weineke was the sheriff.

In Bulpitt, the biggest part of them Lithuanians stayed Progressive. I'd say about eighty-five percent of them stayed Progressive. Along about 1936, 1937, you could just see them migrating, just drifting back to the UMW. You couldn't blame them really. The mistake they made is that they didn't go back in 1933 and get their regular jobs back, instead of letting somebody else come along and take it

Divided Kingdom

Figure 63. Striking miners heating coffee outside Peabody Mine No. 58 in Taylorville, Aug. 20, 1932. On September 1, 1932, the strikers organized the Progressive Miners of America to combat Peabody Coal and the United Mine Workers. (Courtesy of *Decatur Herald and Review*)

away. Really, there was no way they could win that war. It was a loss and that's all there was to it, and it cost a lot of lives. People were getting in trouble over nothing. Just like when they bombed that house in Tovey two doors next to us. The man was a Peabody thug and he come charging out of there blaming everybody for bombing his house. He was blaming innocent people. That's just the way I saw a lot of things that happened.

Ray Tombazzi was one of the tough Progressives, him and Eddie Newman. They had another guy by the name of Roger Tietran in Tovey. We had one of the toughest men in the country right here in Bulpitt, John Gibson, who eventually ended up in Washington, D.C. They [Gibson and Tombazzi] wouldn't back down for nothing. Some of the UMW guys tried to run Tombazzi out of Kincaid one time when he was riding around the square. He [Tombazzi] got out of the car and pulled two 38's and stuck them under their noses. "Now, which one of you two guys wants to run me out of town?" Neither one of them opened their mouths. "Now both of you guys get out of town." They took off on high. That's the kind of stuff that went on.

One time Eddie Newman stopped my dad up there on Renee's Corner [a drugstore-lunchroom in Taylorville]. Eddie Newman and my dad were the best of friends. My dad had an old 1928 Chevy. He said, "Pete, would you do me a favor?" My dad said, "Well, sure, Eddie, you know I'll do you a favor, whatever you want. What do you want from me?" He said, "I want you to take me and them two swampies I got sitting there on the curb out in the country. I'm going to beat the hell out of them." My dad started laughing and said, "I can't do that. I'll do anything else. You want some money, I'll give you some money." "No. Pete," he said, "I wanted you to take me out in the country. I'm going to whip these two guys." That's the kind of stuff that went on.

Alvin Amuel Wise

Alvin Wise's poignant story graphically recounts the hardships under which PMA miners labored.

After being married, we came back to my dad's house. The men who had gone on strike with the Progressives were sort of going back to work [with] the United Mine Workers. My dad went back to work. He had to go back to work; there was no way out of it, or starve his family to death; he had to go. They were eating a big pot of oatmeal for one meal, like for breakfast, that was all. Maybe they at a pot of soup for supper; maybe they didn't get no dinner. I was sort of heartbroken when he went back to work. We thought that we were fighting for a cause at the time, but we found out later that the cause was just starving us, really. This was in 1934. He held out for about two years and he went back to work.

I went back to work in 1936. When the trouble began, I was in Christian County for awhile. My dad went back to work and he sent me up to Detroit, Michigan. I stayed up there six, eight months in 1934 with an aunt of mine so I wouldn't get in any trouble down here.

I never did get work; I couldn't get a job. I tried. My uncle worked at Chrysler up there. I couldn't get no job at Chrysler; he couldn't even get me a job. I went up there and stood in them lines and tried to get a job. Then I came back from Detroit and I started staying at my mom and dad's house.

After they moved out to Cowden, Illinois, I went out there. The Progressive miners had dwindled me down to one beat-up pair of overalls and one shirt.

Do you know what we lived on? Rabbit and cornbread. I got the rabbit fever [tularemia] out there. It took almost four years to get over that. They had to operate on me under the arm after I had it for about six months. I was down there below Cowden when I caught the rabbit fever, that is from skinning rabbits. I had to go out early in the morning and shoot a rabbit before you eat your breakfast. But we ate rabbits, boiled turnips, and cornbread, whatever we could get; that's what we ate. I was doing the Progressive miners.

My grandfather and my aunt went all around making speeches for the Progressives and everything. His name was Jack Spinner. He was an organizer at the time. They would go around making speeches about the conditions and the reason that they came out on strike.

There was no big shots as far as the Progressive miners was concerned in this community. It was just people that was out on strike for a cause. At first, we thought the cause

Divided Kingdom

was good, and I still think the cause was good. There was very few of us that ever got helped by joining the Progressives. We never did join the Progressive miners like writing your name down.

My aunt was the head of the Women's Auxiliary: Laura Spinner, nee Laura Clerek. She just went out with the guys—with Dad—making speeches, just like he did.

John Bellaver

Progressive miners from other counties south and west of Christian County came to the aid of their striking co-unionists in picketing the Peabody mines in 1932 and 1933. John Bellaver and his family joined the picketing in Taylorville in the fall of 1932, and traveled to and from their home near Hillsboro where they were working in a PMA-affiliated mine.

I was here one night in 1932 and the telephone rang. They [the Progressives] were trying to get the Peabody mines closed up in Taylorville, see. They called and said, "Hey, everybody go to Taylorville tonight. All you Progressive miners."

I just moved into this house [in Schram City, Montgomery County] about 1932, I think it was. My uncle had the car. I couldn't afford a car. Me and him went up there. You know who we went and seen? My wife's uncle. He was a scab. He had a bunch of scabs from West Virginia in his house and some from southern Illinois and Kentucky. I went in there and when I looked out amongst them I said, "We hit a hornet's nest here." I just kept still because, hell fire, you could look at their mugs. That was enough for me. Bunch of thugs was all they was. Peabody was paying them and he was boarding them. And he was a boss out there at No. 9 where I worked.

Divided Kingdom

We didn't get very far with them thugs and scabs. They'd chase you out with guns. The state police and militia was there. They knew we were coming. They had more militia than they had miners. They'd chase you back to the Christian County line. The Progressives was going to try to [shut the Peerless Mine down in Springfield]. A guy told them, "You'd better not go there. All they've got there is gunmen. They've got tons of them out there. They're just waiting for you. They'll kill every damn one of you guys. They did kill a guy there on the street. Edris Mabie, wasn't it?

The women's auxiliaries were active. They had more pep and more guts than the men. If it hadn't have been for the women they wouldn't have had nothing. The women were smarter. They marched and weren't scared of nobody. They knew them state militia wouldn't bother them. They wasn't ascared of the militia.

The sheriff and his gang caught a bunch of us miners from outside the county and they put us up in the courthouse in Taylorville. [The Progressive prisoners] fixed that courthouse. It cost them at Christian County that day. The miners tore every book they had; they just tore it all up and throwed it at them. They said, "We'll take care of Christian County for you.

Oh, there were four mines there in the Taylorville area. They all had about eight to nine hundred men working there. Oh, there was a lot of scabs there. I worked in Taylorville and I know. It was a depression pretty heavy, you know, in them days. And people were scared for their jobs. And them, miners-they'll work. The boss would make them do anything. Anything.

In Kincaid they had a No. 7 mine shaft used to come up and the men used to walk out of the cages this way right north, see. I was there picketing when these scabs would come up out of the mine. The striking miners were sitting over in these houses, you know, right across about two, three hundred feet away from the cage when it come up. A lot of them boys didn't get off that cage. Vrooooom. Machine guns! "You God damn Kentuckians!" The striking miners said, "You know what we done with them?" I said, "Hell, I don't know." They said, "They [the scabs] are buried there in the lot on the coal property." Boy, I'll tell you, it was dangerous.

Peabody finally broke it up. Well, the scabs is no good to them. They was just costing them a fortune. They wasn't getting no coal out. So Peabody finally broke the strike up, by rehiring a lot of the striking miners back. They [the PMA miners] had to go back as UMW because they needed the money. When you're broke, you do things. A lot of them guys didn't own their own homes. They were paying rent.

Strikebreakers

John Wittka

Miners who stayed with Peabody Coal and the UMW felt the full force of community fury. John Wittka's family lived in Langleyville, where only two other families continued to work at the UMW mines.

We were strikebreakers in the little town of Langleyville while it was good and hot. Ha, you should never live through that. That was horrible. Your best friends turn against you. They called you names, cussed you. They beat up my dad and my uncle. My dad's face was like that [puts hand against face]. Oh, I tell you I will never forget that. Then when they all seen we was making money and nobody else had a job—Progressives weren't working—they all started drifting one at a time to the mine. Back to the UMW.

There was only three Germans in this town of Langley; we lived close together. One here, here and here [all next to one another]. One guy lived just east of us was a bachelor. He was a big guy named Ed Hencha. He is gone. My uncle, his wife and their boy lived right across the street and that is all there was. We all went to work. It was hot. We had the whole rest of Langleyville after us—Italian, French, Slavic,

Lithuanian. Most were Italian. I would say forty-five to fifty percent. Nobody wanted to be first to go back to work. Somebody had to break it; we did it. My dad seen a chance for me to go to work. That is why I got a job at seventeen.

They harassed us, too. We had to lay out in the front yard, one in the front and one in the back, with a shotgun at night. They would throw rocks at our house. Throw rocks at anything they could see. It was plain hell.

When we come to Taylorville that was all right, because there was lots of people working in the mine, too. So I went to and from work in the militia truck. It wasn't even safe to walk to work. We rode in the militia truck until some of those strikers went back to work from Langley.

The strikebreakers from southern Illinois were Italians, Americans, Kentuckians—they were a mixed breed of people. They came from all the European countries over there. Most of them were miners. A lot of them were good workers; every one of them was a good worker. That's why they all congregated together.

All those divisional superintendents at Peabody had their henchmen. That's the whole thing: to break the PMA down and to protect the United Mine Worker's union. We had this happen right along with mechanization. They had to have a strike to get rid of so many men. So they come along with the Progressive trouble and some other strike to eliminate some more men. The machines agitated the men; they organized and then they'd strike. Boom! The company only takes back a few men—no job share, nothing. One loading machine could replace twenty to thirty men.

Lena Dougherty

Lena Dougherty shows how common it was during the mine wars for friends to turn on families who remained loyal

Divided Kingdom

to the UMW, and how the threat of violence hung around these families.

The day that I moved out, we were leaving the store and here was all my friends with clubs and stones; they were going to stone us. If you think that didn't hurt. We loved them people. I couldn't believe they would do that to us. But my husband said he wouldn't go. "They are not going to chase me out." But, of course, he couldn't stay behind in the store.

One incident that I never will forget. One morning just before we left Tovey, my sister came and she was as white as a sheet. She said, "My God, where is Bullo?" We called my husband Bullo, that was his nickname. I said, "He is in back." She said, "Oh, my God." I said, "Why?" She said, "When Joe [that was her husband] got up this morning he said, 'Well, I want to tell you now. I couldn't have told you last night,'" And she said, "What?" He said, "Your sister's husband will be hanging from a tree right now." She said, "You have to be kidding." "No, they had it all made up; they were going to lynch him and hang him from a tree. I wouldn't tell you, because I knew if I told you that you would be upset and you would go down there or something. We can't get into it; we are in business, you know." And so I said, "No!" But that's what the plans had been; they were going to hang him. And this was before he started working. He wasn't working at the mine. We had lived in Tovey from 1927 until 1933. So that's six years we had been living there when this happened.

It was a tough decision to stay with the mine. He said, "Lena, I have got to; we need the money." Miners couldn't save anything, because there would come a strike and you would lose what you had saved till then. You could always depend on that. Contracts were signed for two years. You always had to save toward April. And you didn't save just

207

by dollars: you saved by nickels and dimes to have a little nest egg. If you had two hundred dollars, you were lucky. How long would two hundred dollars last back then?

Then my husband went over to Langley to work. And there was about ten from Tovey that wouldn't join the Progressives, so, of course, the sheriff deputized them—those ten. You had to carry a gun for your own protection. This house was fired into during the mine trouble, but nobody knows what side did it. The Progressives will say the Mine Workers did it, and the Mine Workers will say the Progressives did it.

Have you heard about when they bombed Fred Eddy's house in Tovey? Fred Eddy was an engineer for Peabody Coal at the mine. We were just two doors away from Eddy's when they were bombed. I wasn't home when this happened. My mother-in-law and my father-in-law—who lived with us—were with the kids when that happened.

My husband said afterwards, when it was all over with, "I will never go through that again. I would take you someplace else." But what could he do? All he knew was the mine. The one year they were on strike, he went to Chicago. He worked in a packing plant and stayed with his aunt. Then, when the mine started working, he came back. All he knew was the mines. Besides my husband, miners who stayed with the UMW were the Roberts, Lisses, Eddys, O'Connells, and some others I can't remember.

My oldest daughter was in the second grade when we lived in Kincaid. She said, "Well, I remember they used to call me scab." They couldn't go on the street alone; they had to be with two or three others. When we got to Kincaid there were quite a few mine workers employed—about twenty families. But in Tovey, where we had just left, there wasn't that many—maybe about ten.

Out-of-work miners were afraid to join the Mine Workers. You take the people we used to live catty-corner from. They were Lithuanian; they could hardly talk American. Here in Bulpitt, they were one of the first to break with the PMA. He was Joe Makenas; he was one of the first that went. But they were afraid to go to the mine; they were afraid for their lives. Mrs. Makenas said, "When Joe started working at the mine we were so afraid he'd get killed." It was nasty, it was bad.

They didn't hurt my husband because Bullo was outspoken. If he had something to say, he would say it. He had a gun in the car. He didn't have to have it to go down in the mine, but had to have it in the car because when they worked in No. 8, they use to come through what they called the bottoms, over there by Jerseyville. They were so afraid when they would get down in them bottoms that they would be ambushed; it was bad. You don't want to think about those things, but that always hurt me so bad; those were our friends.

Merle Ahlberg

Merle Ahlberg relates how the PMA-UMW conflict bitterly divided families, pitting brother vs. brother, father against sons.

My dad had a family to raise, and he undoubtedly thought it was right [not to strike], because that's why he stayed at work. He had a brother that went the other way. Until they died, they never talked no more or anything. Ours wasn't the only family that happened to. There were so many families got split, as you'd call it, because some of them could see one side and others see the other. Naturally, I thought my dad was right. He wanted us to have food to eat.

There was never no hard words between dad and his brother. They just didn't push each other. One of them didn't say, "How come you kept on working?" and the other one say,

Divided Kingdom

"Well, how come you quit and went to the Progressives?" There was none of that. My mother visited with my aunt and us kids played with each other. I was twelve-thirteen-years-old at that time.

One time, we were up on the square and we had gone to take the women to the grocery store. I was carrying a couple of sacks of groceries and dad was carrying a couple of sacks; that's the way we bought, a week at a time. There was never no charging groceries or candy. We were coming in front of the bowling alley. Along in front was a little curb and men were sitting there. They were all Progressives, and, of course, when we went by there were some murmured remarks. One of the Progressives got up—and I won't mention names—and turned to the bunch of them sitting there on the curb, and said, "Hey, lay off my friends." That stopped everything. Dad was well-known and he was well-liked; he wouldn't run from nobody.

I thought about that so many times. Later I found out why that happened. This young fellow was brought into the coal mine by his dad and, of course, back then you had to have somebody to sign up for you when you went down below. A father couldn't sign up for his own boy. So this old man come to my dad and said, "I want you to sign up so-and-so so he can get the job." So my dad went ahead and signed and took him as a buddy when he was hand-loading coal. This boy was mischievous. He liked to fight and he liked to wrestle; he liked to have fun and was strong. One day he was working alongside my dad and dad told him to come out of a place that was bad and this boy didn't understand anything like that, so dad told him again, "You get out of here." He still didn't pay no attention, so my dad went after him and just literally threw him out of the place. This guy was going to beat up my dad for doing that, but about then the roof come down, and that's where he had made a friend. This boy went with the Progressives, and, of course, dad stayed with the UMW.

Even in the neighborhood of say, fifteen or twenty houses, there would be both sides represented. John Wittka stayed with the UMW. His dad worked in the mine too. Armstrongs, from out in Hewittville, and Binghams. All had big families. We all lived outside the city limits. And a lot of these UMW men were from City Park area and Hewittville area. Hewittville is where mine No. 58 was. My dad's brother-in-law was UMW; and Joe Berrington-in later years was one of the deputy sheriffs of the county-he was UMW of A. Mr. Wilson, I can't think of his first name right now, he was another one. These men stayed together in a group. There was only two ways of getting to the mine, other than if you had your own car you either rode the streetcar or you walked. Mine 58 was a little over a mile from where we lived, so these men, a lot of times, would walk when they was coming home from work and they would talk and they would joke about stuff going on.

Kenneth Cox

Kenneth Cox moved to Taylorville from the southern Illinois mines. Miners who broke the strike came from throughout the Midwest to work at the Peabody mines. All of the miners—especially those on strike—were "tough customers."

I brought my brother up here in 1932, October; he was a coal miner in southern Illinois, and he worked at Old Squirrel Coal Mine, which had shut down. The bosses gave him a letter to bring up here to get a job. So me and a friend of mine brought him up here to go to work. I didn't intend to go to work when I came here, but when he came out of Wardy Argust's office, the district superintendent, he had a slip for me and him both to go to No. 9 and go to work.

Divided Kingdom

I said, "Well I'm not about to go to work in the coal mine." He said, "Yes, you are. You need a job and you are going to stay." So I took a job on top, and I was on top for about three years before I went below. That's how come I came to be here in the coal mine, on account of him.

We went to No. 9 coal mine and Bill Starks was superintendent out there, and he talked to Floyd, and Floyd took us in as a couple of boarders. There were two other boarders. There was four of us. All four of us was from southern Illinois. So we stayed there. Cy Cox was married, and he finally brought his wife up here. They moved out on North Webster Street, and I boarded with them out there on North Webster Street until I got married. They apparently had just reopened No. 9 because they were taking on hands—you know, they was hiring people.

In Mine No. 9 there was some of them from the Taylorville area and some of them from out of the Taylorville area. We had people in here from Indiana; we had people in here from Kansas. Bill Ridley was a mine manager at one time. Now, he came from Kansas originally, and there was quite a few of them up here from Kansas. There was two Menson boys and one of them was the assistant mine manager at No. 9. They came from Kansas to here. There was people here from southern Illinois that had worked in the coal mines down there; the coal mines had worked out and they was closed. They got a job here and went to work here.

The UMW originally organized the coal miners around here. They fought for it. Some of those coal miners years ago [1898] lost their lives over this, you know. In fact, we've got a monument right out here in the cemetery in memory of these guys. They actually died to organize the coal fields in this area. It wasn't only true of this area, it was true with coal mines in Kentucky, Tennessee, and Virginia. They'd send organizers to the coal mines and they'd find the organizer in the bushes with a bullet hole between his eyes, you

Divided Kingdom

know. So the United Mine Workers is the original union that organized the coal fields. When we went to work under the banner of the United Mine Workers in 1932 we were called scabs. I don't understand it; it is beyond me.

Oh, we had our problems as so-called scabs. Now, the wife and I moved out on the Vanderfer Street and the next door neighbors—Harlon Danny and his family—were Progressives. But they never did do anything wrong to us. We was always friends to them. Some of their people, somebody died there, one of their family, and my wife baked a bunch of stuff. I took it over, pecked on the back door and they came to the door. I gave it to them as a neighbor and they took it. He told me later, "You know," he said, "that kind of got next to me." That was probably in 1934.

Of course, some of them wouldn't talk to you. On Christmas Eve, the wife and I was uptown doing some shopping and I sent her down to a store to get some oranges. I was standing up on the corner waiting for her, and between where I was standing and where she went was a tavern. They called it the Brass Rail, and it was a hangout for the Progressive miners. They'd get in there and when you'd come by, they'd call you scab, and a fight would ensue. I mean a gun battle. That night they had one guy out in the street; he was the head of the United Mine Workers here, and they was shooting at him. Of course, he was shooting back. The guy was using a high-powered big gun, and the guy that shot back at the Progressive was shooting a small .25 automatic. But he was out there in the street, jumping back and forth and that guy was shooting at him. Another guy walked right in front of them and he got shot. My wife, she come through there just before the shooting started. When that shooting started, that scared me to death, you know.

Of course, we did have some guys come in here from around Chicago that went to work for the coal mines and United Mine Workers. We're pretty sure that one of them was a gangster, because him and two fellows went over to rob the Mount Auburn State Bank. The bank had already been notified by the UMW they was coming. They wanted to get rid of this guy, see. He didn't even attempt to rob the bank, and they chased him down in the corn field and they killed him. To me, there was thugs on both sides of the fence as far as that's concerned, but this is the one I knew.

You don't go out and shoot at somebody unless you're about a half thug. But when we first came up here, Peabody Coal Company sold some guns to some of the coal miners. I was riding with a bunch of fellows and I'd take three and four guns every morning in and leave them in the commissary where you bought your stuff, and they'd put you name on them. Then you'd pick them up that night.

Paul "Stormy" Dixon

Dixon migrated to Taylorville from the southern Illinois mines, which were working irregularly. Dixon estimates that nearly half of the miners working in Peabody's Christian County mines were from the same southern Illinois mines in which he had worked.

It was about 1928 they put the loading machines down in southern Illinois. When they put the loading machines in, they also put in conveyors; they did away with the hand loading and went into the mechanized operation. They had to lay off many of the men, because they didn't need that many men to run the machinery. Well, they didn't know what to do with these men, so they had a division of work. They classified each man as a motorman, a trip rider, timberman, a conveyorman, or whatever. Then you worked your share. You work a week, and then maybe you're off three weeks. They would let you work until you got six days in. And you

Divided Kingdom

had to be off to let the other men in that category work. Sometimes, if there wasn't very many men, you got to work pretty regularly. If there was a lot in your classification, you had to wait. If you worked three days a week, it took two weeks to get six days, so you were off eighteen working days. It would be six weeks before you worked again.

The workers didn't like that idea of job sharing, but they had to do it. They didn't have any choice and, of course, they would try to get jobs other places. The local union had the idea about job sharing. That's what caused all this trouble with the union—the machines and the unemployment. They had a White Card Union [UMW] at one time, and later on they went into the Progressive Union. We didn't even elect our own state president under the UMW. They said we couldn't handle our own business here.

In 1933 we weren't working very good in West Frankfort. I was talking to Superintendent Fred Burnett about the mines and how they were working, and he told me to come up here: that I would work weeks up here, where I wouldn't work days down there.

Burnett gave me a letter to come here to see Superintendent Argust. I got transferred up here to Peabody No. 8 to break in as a driller. They were just starting the loading machines in 1933 at No. 8. They only had two machines and nobody really knew how to run them. They didn't have anybody that knew how to drill coal for them and I started there drilling. I had to shoot up all the bottom that the coal loader had left.

I don't know how you would word this: they took their favorite miner and they tried to make operators out of them. Some made good operators, and others never did. Then later on they started a third and a fourth machine, and after they got other machines the men came from other Peabody mines. They came from Danville and they came from southern Illinois. See, southern Illinois, about five years before they had them out here at No. 8, there were people that ran machines down there, and they knew more about how to maneuver them and handle them and dig the coal down.

We had trouble getting machine operators because we had that mine war and a lot of the guys went Progressive; they wouldn't go to work. The Progressives were against the UMW-A. Peabody wanted the United Mine Workers to stay and they paid most of the bills for the guards and the fence around their property. They didn't want the Progressive Union. Their contract was with the UMW-A.

A lot from West Frankfort heard about the works of the Peabody Coal Company. It is just like anyplace. If you are having a boom somewhere, you know about it. A lot of them come up on their own. Peabody No. 18 [in West Frankfort] was still working when I was sent here.

I would say about half of the strikebreaking miners in Peabody Mines in Taylorville or more were from the southern Illinois area.

Otto Klein

Otto Klein came from the southern Illinois mines of Sesser where he had been unemployed. This particular narrative illustrates how many of the imported UMW strikebreakers felt toward their PMA counterparts, and why they were willing to fill employment vacancies.

I actually sympathized with the Progressive miners. We didn't come from Sesser or southern Illinois to break the union or take their jobs. We came here mostly because we were trying to survive. We were out of work. The grocery store owner was threatening to shut us off. It came to the time where either you go to work someplace, in

Divided Kingdom

the coal mine, or steal, or whatever. So, actually, I don't think that the miners from southern Illinois came here to break a union, or to be strikebreakers or anything like that. We actually sympathized with what they were doing.

There must've been ten or fifteen strikebreakers that I know of that came from southern Illinois. There was some here when I got here. Evidently, there was different ones from different areas, like around Ziegler, or West Frankfort, and maybe Harrisburg. I would say probably fifty percent of miners working the mines when the strike began were from elsewhere outside this area. See, we left southern Illinois because of loading machines!

In the southern Illinois mines the miners negotiated, and the company said, "Okay, if you don't let us put down the loader, how about putting down the conveyors?" They went through a meeting and found out that if they would split the work—job share—what it would do would give guys one week on and one week off. They figured that would be the best. Otherwise, they wouldn't open the mine. In the future, the company would probably shove them Joy loaders down there anyhow, and then it would've been a real squabble amongst the men.

See, if they put in loading machines, there aren't going to be very many left working; six hundred eighty-five guys with about ten loading machines. That's about maybe one hundred eighty-five working and there would be about five hundred off. So we decided that we would go along with the belt loaders—conveyors they called them. There was two kinds: there was two wheelers and there were four wheelers. Four wheelers stayed on the track but they was mostly for entry work*. The two wheelers you could throw off the track, swing them into corners of the room and load coal that way. That lasted from 1928 to 1930. This was an alternative that they had to accept in 1928. After that, it was all loading machines.

Divided Kingdom

When we came here from Sesser we knew of the threats and the bombings. My sister got a letter from someone to tell me and my brother where to go. So when I got here I went to a house on Cherokee Street. I took the letter to the house, and I guess she was a little afraid of getting bombed. She said, "No, we don't take boarders." She directed us to Reverend Odam on Poplar Street. There were four miners, and at different times there might've been five at Odam's.

We were boarders. We had the rooms to sleep, and he did the cooking. Me and my brother-in-law stayed there, and there was about two other guys that boarded there, too. One of them must have been a Peabody hireling, because at supper time—he did it maybe for show, I don't know—but he took a revolver out and laid it on the table while he was eating supper.

So when I got here, I had a letter to get the job from the big superintendent in Marion, and we took it to the secretary at Peabody. All the letters that came in go through her and they wind up on W.C. Argust's desk. Okay, then we go back out into a room and we all sit around and wait for someone to take you back into the room. I think it was Blackie Vickery that was the guard. He would call out a name. The guy would come in and then have a short interview, and if he came out smiling, that meant he had a job, I guess. So then there was a couple of guys that didn't have a letter. Pretty soon Vickery came out and said, "Anybody that doesn't have a letter, the hiring is over." So me and my brother-in-law just sat there. Blackie didn't know we had a letter. Pretty soon there was some guy called him over to the side, and he done some whispering and whispering and whispering, and I think he bought himself a job, really, just to be truthful. So next thing I know, he's in the office and he comes out with a job. I knew him, because he was from Buckner, too. Then just me and my brother-in-law are left. Blackie come out and said, "You guys got a letter?" We said,

"Yes, we have." He said, "Why, I didn't know that." So we went in and we both got a job at 58.

I got a job as third man on a loading machine working for W.C. Argust's son. It was easy to get fired at that time, because the pit committee was a friend of the companies. In times like this, there were guys out at the gate trying to get in. There was guys up at the office without letters that couldn't get jobs. Jobs was at a premium, and, in fact, they even got guys paroled from the prison to work at the mines. Different ones at different times because, I suppose, there was some rough goings-on and these guys was used to that kind of stuff. Coal miners normally have their problems with coal companies on picket duty or something like that. But when it comes to coal miners versus coal miners, it's hard. You don't look at a Progressive miner and hate him or anything. He's a coal miner, you know; that's the way it was at that time. But if you get a guy that's come out of prison on parole or something, he doesn't care who he's looking at. So there's a difference between those kind of miners and regular miners, like we were.

Perry Gilpin

Perry Gilpin was working on a farm when news of the strike reached him in 1932. The wages Peabody offered were a strong enticement for poorly paid farm workers.

This was in the fall of 1932. I was working with Mengleson's Box Company on the farm for ten cents an hour doing everything. That was the worst job I ever had. I was vaccinating chickens. But I read in the paper that night that they were hiring men at the coal mines for $2.85 a day. Man, that is big money, right or wrong. I had a little black suitcase, about that long and that wide, [10" x 20"]. I

Divided Kingdom

just took a pair of coveralls with me, and I headed down the road for home. I went down to old Bill Hardy's house in Taylorville and asked him for a job. But it wasn't as easy to get a job as I thought it would be. You couldn't even get down the street or to the mines for the pickets. The roads were so full. "Well," he said, "If you are scared to go out to that mine, you will be scared to come to work." So, the next morning I fought my way through. 'Course, they had militia all around Mine No. 58. I got over there and old Bill Hardy came out and hired me. Well, there was three or four of us then—Levi Wilson, Alex Dougherty, Woody Richard, and Charlie and me. So, it wasn't too bad coming home that evening. We all lived right around one another. But going out there in the morning was when you had a rough time getting through. You didn't know when one of them was going to knock you in the head.

We had been uptown one night and come home about midnight. I just got in bed and they blew down Tombazzi's porch off the house. Broke windows in our house and the old man thought, by God, they done bombed us 'cause I was working.

I went through the gate by myself with pickets on either side. They was right there at the gate. Bill Hardy come out and looked them over and the ones that talked to him he told them to come on in, see. Old man McGuire—Ike McGuire—was a gate watchman. He was a Johnny Bull or something, you couldn't understand him. I know when he opened the gate I was going in, anyway. I wasn't going to fight them pickets going back. Harry Renals had a sign in there. He was PMA. "Don't take jobs from your brothers and starve our babies." And he didn't have a kid. But I laughed and that militia guy stabbed me in the back with a bayonet and told me to get going.

The PMA caught me on one night up on the square. I used to take a four-ten shotgun. I'd take that four-ten there and take it apart, see, and then I would hide it under the steps of the Old Freemasters Church and go on to town. They picked me up one night on the square; I had them shells in my pocket. I didn't know how to get out of that, but I finally wiggled out of it. They kept my shells. I had the gun then, but I didn't have no shells.

John Ralph Sexson

John Sexson's story is a pretty grim reminder that even those miners who stayed with the UMW were pressured and intimidated by the coal company. There were always hungry and out-of-work southern Illinois miners at the gate waiting to take any work that became available.

I would say one third of the business people of Kincaid sympathized with the Progressives. Ben Tex's store [a UMW sympathizer] was shot up one time, if I remember right. I don't remember who done it. On either side you got certain men that want an excuse to got out and blow this or that up. I don't think I'm wrong.

In Taylorville there was bombs! I would here them bombs go off and I would be at home and wonder whether I was going to go to work the next morning or what was going to happen. I never want to go through a thing like that again.

In 1932 you could easily get beat over the head, or whatever, if you talked to the wrong person. That's the reason the militia come down there. They finally brought the militia down and two was a crowd up on the Taylorville square. They wouldn't let them congregate. Anymore than that they would break the crowd up. They arrested very few. They would just simply break up these things before they got started, or they tried to. And as far as the bombing, I don't know if they ever found anybody guilty that I know of.

Divided Kingdom

Figure 64. Over 5,000 striking miners surrounded Peabody Mine No. 58 in August, 1932. The militia dispersed the pickets but they reformed at the mine gates throughout the fall of 1932. (Courtesy of *Decatur Herald and Review*)

Along in there the C&IM [Chicago and Illinois Midland Railroad] had guards at the bridges and stuff like that. Well, they got the wise idea to check us off; I think it was a dollar a day to pay for these guards. Well, we rebelled. Some of us rebelled. I was one of the ringleaders who rebelled in that situation. I couldn't see paying the C&IM company for these guards they had on the railroad.

I understood if you had ten or more names on a petition you could call a special meeting of the union. We got the signers and gave it to those pit committeemen on a Friday. Monday morning I got word that the superintendent wanted to see me down at the office. John Abrell. He said, "I don't understand what you guys are trying to do. We got two unions here now. We got the Progressives, we got the United Mine Workers. You guys are trying to start another union. We ain't going to allow this at all." I said, "Well, I don't like paying for these guards. We're not going to do that." He said, "I don't know what you are going to do about it. That's the way we're going to do it." I said, "Well, if that's the case, I'll just drop it and go ahead and pay the dollar." I think we did, for awhile.

Abrell got you scared to death you was going to lose your job. He was a good talker and I couldn't hardly win. I tried, but I did no good. I don't know whether we had that checked off a dollar a day, or a dollar a week. It don't look like much, but to us at that time it was quite a bit when your salary was $4.75 a day as a motorman, and you're trying to raise a family and pay rent.

Well, this Progressive break didn't happen all of a sudden; this just didn't come up overnight. See, they tried to form the National Miner's Union, I think, in 1927 or 1928. That was another break away, but it didn't last long. The next to come along was the Progressives. They tried to form a union and break the United Mine Workers, and they had no luck the same as the National Union. Some of the same element that was in the National was in this Progressive movement.

We had pickets in 1932. I mean pickets out of this world all the time and bombings around and in Taylorville! The Progressives would dynamite buildings. They'd keep telling us there was going to be pickets tonight, and I was down below and then would come up on top and, nobody there. So, finally one night at No. 7, [January 3, 1933] we come up on top and they was handing out pick handles. It was one of these picks with the pick handle off, you see, and putting a rope around, so you could carry them in your hand like a club.

I thought there was going to be a knock-down-drag-out. I went on out across the mine track, and boy, I heard this popping and cracking and I thought them was blanks they were shooting. I finally got myself behind a little tree, and I mean I was squeezed up behind that. There was, by the way, this boy; he was just a boy, probably eighteen or nine-

Divided Kingdom

teen. It was the second week he had worked at the coal mine. I wish I could remember his name. He got shot and killed. There was three of them that got shot and killed that night. I remember him so well because he was standing there right next to me, you see. I don't know if I worked the next day or not. They finally got things quieted down and I got out of the mine yard and got home.

I lived in Taylorville. You know from Taylorville to Kincaid you got a lot of hills and timber and stuff like that. I was scared to death to go through there. I will tell you; I'll admit it. They [the militia] would bring me out and bring me back home at night to Taylorville. And I done that for quite a bit—rode back and forth with the militia. I was afraid I would get shot going to work. A police car is what it was. You could go up on the square and ask them to take you out to work, and they would take you. You didn't know who to talk to while you was on the square. You could get in a fight pretty quick.

The Progressives shut down No. 7 for about two weeks. No. 8 didn't start up, if I remember right, for about three or four months after the big mine war in 1932. The war started in the fall of 1932. You didn't know how bad it was. Being a United Mine Worker, I didn't go up and drift around town too much. I didn't want to get in any bouts or anything; I didn't have to, see. One fellow who lived down in Taylorville tried to talk me out of working. He said, "You ought to join the Progressive Miners. You're not going anywhere where you are." "Well, I don't know. I think I am satisfied the way it is."

They'd [Peabody] hire them from anyplace. If they wanted to come in and ask for a job, they would put them on a payroll. They had to see the division superintendent, who was in Taylorville at that time, to get a job. He would designate which mine you were to go to—7, 8, 9, or 58. Taylorville was split between PMA and the UMW—but Kincaid, Tovey, and Langleyville were pretty radical for Progressives. I mean, it was a pretty tense situation for quite some time. It was brother against brother, or brother against father; it made no difference.

It was really important to be working then. I don't know how, looking back, we even survived with the low wages. Of course, everything was low, too, you see. I started work at the coal mine at $3.00 a day and then went down to the local for $4.75. That was a good raise for me, but after the first year they slacked off, and after the strike two or three days a week is all you worked. That was 1937 or 1938, I believe.

Back in those days those Progressives really had a lot of people. They had—I read in *The Progressive Miner*—one hundred and six locals back in the beginning of 1933. I believe, all through the Benld area and down through there, they all went Progressive, and some around Springfield.

I wondered at that time how they was ever going to get the thing off the ground enough to form a union. Which they never did; I mean, compared to the United Mine Workers. I am well pleased, myself, the way the things turned out.

We at the UMW always took the stand-or I did-that we was all going to starve to death together under the job share of the PMA. As far as I was concerned, I would rather work steady, or not work at all.

Donald VanHooser

Businesses that sided with the PMA experienced financial difficulties and most eventually failed, as VanHooser recounts. VanHooser was also a pro-UMW miner who personally felt the anger of PMA sympathizers which he relates in this narrative.

With layoffs the men knew ahead of time what they were going to get. Somebody had to be laid off because the machines were taking their place.

Divided Kingdom

Some of them got jobs elsewhere; some of them got out of the mines altogether. It worked out alright. See, John L. Lewis was our international president, and he warned the men when these loading machines were coming in that they were going to take the place of men and there would be a reduction in wages. The union was prepared for it.

Now, when we had that mine trouble, that was something else. It was 1929, the first time, and Andy Fletcher was the sheriff here in Christian County. He sent a deputy to us at our boarding house. The deputy said that Andy wanted us to come over to the courthouse. We went over there and he said that he wanted to deputize us so that he could have more deputies, which he was authorized to do to take care of some of these problems with the striking miners. I didn't like the idea, but I said, "Well, let me go get some advice." So I went to a lawyer, and he said, "You might as well take it because he can put you in jail. You can't refuse, not unless you're sick." So I was deputized.

The strike lasted only three weeks the first time. We had militia in here for three weeks at that time. We had Company C, the cavalry out of Springfield. Company B was from Decatur.

Next were the Progressives. As a deputy, I had to go to Nokomis to get some people that was arrested and bring them back to Taylorville. We had to take the C&IM train to Kincaid and pick up twenty-five people that had been arrested for demonstrating in the fall of 1932. We brought them back to Taylorville and they were processed out at the courthouse. Some of them were put in jail, and others weren't.

Later in 1932, John [a brother] was working at Gebhart's store down in Centralia, and I was working here. When the mines went out, I didn't work. John was up here over the weekend and he was with me at the laundry—that's where I had a room. They [Progressives] caught us on the corner east of Main Street. There was about fifty men that were out on strike, and they caught us and they kind of worked us over (laughs). We finally got away from them and got across the street. I then just said to the gang—I knew most of them—"Okay, you've showed your hand. I'll show mine. When these mines go back to work, I'm going back to work. You try to stop me then." We weren't deputized at the time; this was after that first time. And when No. 9 went back to work, I was on the loading machine.

I went back because I was still a United Mine Worker. That's who I was for—not your Progressive miners. I thought that the United Mine Workers would prevail over the Progressive miners. So when No. 9 opened up and they called me back and said, "You want to go to work?" I said, "Yes." Then they wanted me to go over to No. 7 and help open up over there because they had enough men working at No. 9 to operate. So, that's what we did.

Anyway, we opened up No. 7, and then No. 8 opened up. So that was 7, 8, and 9. Well, 58 was working. That was four mines that was working at that time.

Most of them [Progressives] were from around this area. Now, there was some of the men that never did get back into the mine because the United Mine Workers would not accept them because of their actions. Jim Oseland didn't get back; he had to go to Peoria because he made such demonstrations. I was one of them that he caught down there and worked over, see. And I knew it. I said, "No, you are not getting back into the UMW."

Mom and dad lived out northwest of Edinburg on a small farm in 1930. They got a letter saying that if we two boys didn't quit working in the mine, that they was going to kidnap my sister, Lorraine. To that effect. When Mom brought that letter into John and I, well, it hurt. So I said, "You give me that letter." We went up to Uncle Bill—a Progressive—who lived in the north end and showed him this letter. I told him, "If I quit work, Uncle Bill, somebody's going to get hurt

Divided Kingdom

now. That's all there is to it, because you're not going to hurt Lorraine. I don't think they should've done this." He said, "Let me take care of it."

He must've went up to where the Progressives was meeting on the east side of the square, Boverman's store. The next morning he came and got ahold of me and said, "Don, that's been taken care of. The fellow that wrote that letter has admitted it. We've taken care of him. He is not going to do anything. We'll see to that. So tell your mom and dad not to worry anymore." I said, "Okay."

Another time, brother John had his tonsils out at the old hospital. I left the hospital at five o'clock in the morning and went uptown and was getting breakfast before I went to bed. We wasn't working then. Anyway, I was in there at the restaurant on Main and Walnut-Large's Restaurant-when some of these men saw me and they ganged up. Of course, they wouldn't come in and I wouldn't go out. They finally called the sheriff and notified the militia and the militia came down and made everybody move. The militia had their sidearms, and they had pickets—fence pickets—that they used. That's what they had to protect themselves with. They made the men keep moving, see. They wouldn't let them gang up. There were several hundred Progressives.

We had a number of UMW workers who came here to work. We had a fellow come up from the south, Legs Manasco. He could protect himself. He was good. He was a big fellow, too. Yes, he wasn't afraid of anyone. He was a good friend of ours and he went to work as a strikebreaker. There was a lot of men came up here from southern Illinois and worked in the mines.

Well, I would think it was between fifty and seventy-five percent who were Progressive in Taylorville at first. They were just gradually straggling back to work. They knew that there was nothing more for them otherwise. There was no mine here that was going to go Progressive. They either had to go back and be a United Mine Worker, or else just get out.

In Langleyville, the only person that stayed with the UMW early was John Wittka and his family. The rest of the people in Langleyville, who were Italian, stayed with the Progressives. Even here in town, we had barbers that took sides. I was going to Ray Davis, getting haircuts all the time. I just went in and he cut my hair. But there wasn't a word said between he and I in there, or no one else. When I paid him for my haircut, I just said, "Ray, this is the last one. I'm not coming back." I went to Pete Hill from there on. Pete—he was a United Mine Worker. He was a union man. He and Bob Wright, they were the two barbers. Later, Ray Davis had to close.

The Violence

Frank Borgognoni

Frank Borgognoni was in an excellent position to hear the best of the stories about the Miner's War; his father ran a Progressive tavern in Kincaid where much of the PMA activities were. The violence centered on the real and imagined work of the UMW "enforcers," who were imported as well as local.

When they went on that strike, my father was standing outside of No. 58 gate with a banner when the miners were coming come that night—coming out of the gate. His second man on the machine who still worked, Phil Petty from Taylorville, was sitting in the back seat of this car; he rolled down the back window and he tried to spit on my dad. This is the way he told it to me: he just dropped his banner. He said, "If I've got to stand here and take this, I don't want no part of it." He never did go back. He said, "I'll never go back to the coal mine," and he didn't. That was the end of his coal mine career, during that mine strike.

My father became a Progressive; he ran a tavern down here in Kincaid about half a block. He was the only Progressive miner with a tavern in Christian County. He told

Divided Kingdom

some stories about what went on. One of them I couldn't hardly believe—that the Progressives had a .30 caliber water-cooled machine gun on these buildings. I guess it had to be true. The Progressives needed all the protection they could get; they were outnumbered. The Progressives were standing guard here one night, and a fellow went by in one of these new-type Fords. The first V-8 Ford they ever made, and he was really going down the road. They hollered, "Halt," and he didn't stop. The men opened up and let him have it. Well, they shot him. When he stopped, he ended up on these gas pumps here—dead. He had another fellow with him by the name of Douglas McQuinty who crawled across the field on his hands and knees and got away. He lived. But another time the PMA caught him down here, and a man by the name of Lawrence Store had him on a sidewalk and had this .25 automatic pistol under his chin. He told him he was dead, and my father said, "For God's sake, Lawrence, don't kill him." Two or three guys had to pull him off and they turned him loose.

See, McQuinty was known as a thug—a Peabody thug. He carried a gun for them at that particular time from what I understand. Whether it's true or not, I don't know. I'm just relating the story as it was told to me by the older fellows. I used to sit around and listen to them when I was a boy.

My father was well-respected, and highly regarded as a man with a lot of principle in this county. For a person who was born in Italy, he didn't even talk broken English. They used him as a go-between with the PMA and the UMW at different times. For instance, one time a well known Peabody man—he wasn't a bad fellow, just happened to be on the other side-came up and said, "Pete, can I talk to you for a minute?" He said, "Why sure. Come on in and you can talk to me." He said, "Pete, if you got any friends that wants to go back to work, you tell them they'd better go back to work. It's been two years now they've been on strike. They

Divided Kingdom

have lost this thing." Which was true, they had lost. This was somewhere in about 1934.

So my father said, "Well, I'll talk to some of them." So he went to Tovey and he told them what the man had told him. It was surprising how many people got mad at him and said, "Whose side are you on?" He said, "I'm still on the same side, but I'm just telling you what the man told me." He said, "You do whichever you want to." That's just the kind of a guy he was.

We did need the National Guard, but they didn't do it right when they did come. They were partial towards the UMW of A at that time. But they needed a National Guard to hold down all the trouble. It could've been worse if it hadn't been for them. They had a ten o'clock curfew, and if you were on the streets at ten o'clock, they'd run you in, or they'd take you to headquarters and you were questioned. It was dangerous, people were carrying guns. How bad it was—a carload of these Peabody thugs that was carrying guns would ride down the street and actually grab guys off the sidewalk and haul them out in the country and literally beat their brains in. Then take them back into town and dump them in the ditch. That I know for a fact. I seen one man by the name of Williams, that wasn't taking either side, get a beating. He wore a headband for over three months. His father was a dairy farmer, and he didn't have no part of it, even in the war, but he got his head beat in.

Duke Livesay was the big man for UMW around this part of the country, but he actually never did do any kind of dirty work himself. He had a guy like Fats Orlandi and Andrew Dougherty and a few other fellows that would get out and enforce their demands—whatever they wanted them to do. Just do a lot of dirty work that was all uncalled for, unnecessary.

During the mine wars, if the men would've been smart they would have, after about three years, said "To hell with

Divided Kingdom

Figure 65. Over 5,000 striking miners attended a mass meeting at Peabody Mine No. 9, Kincaid, August 18, 1932.

Figure 66. Funeral for Andrew Gyenes, October 16, 1932. (Courtesy of *Decatur Herald and Review*)

Divided Kingdom

it. We've lost it, let's go back to work." But what they done- what was sad, really sad-they let Peabody import some of the world's worst crumbs that you could ever dream of come to this part of the country and take these people's jobs. There was a lot of good people with the UMW; but, there was a lot of crumbs with them. Most of them came from down around West Frankfort, Benton, and Herrin, and down in that locale. The biggest problem was the ones that they took out of jails and brought them up here. You can imagine what they were. They were animals, really. I mean, it's all over with, but they were animals. They made life miserable, unbearable for all the people around here. They'd abuse them in different ways. If they had a chance, they'd start something, anything.

Mrs. Conkey Engs

Many miners believed that Peabody Coal Company planned sabotage and violence that was blamed on the Progressive miners, as this excerpt relates.

My dad worked in the tipple when the mine wars started in September, 1932. Of course, nobody was working then. One night he came down and walked in with this time sheet to hand to the bosses in there. When he got home that night we were eating dinner and the telephone rang. I got up and answered the phone. "Daddy, somebody wants to talk with you." I said. So he got up and went to the phone and he said, "Sure, of course I won't." That is all he said. When he got back, my mother said, "Well, is everything alright, Grover?" And he said, "Yes honey, I just saw something tonight they don't want me to see."

Later on, he told us that there were three men in the office. He knew that all hell was going to break loose because they were really thugs. He said they were high class thugs, but they were thugs. "It looks to me like they brought in some of the gang to help break the strike up"—that's what he said later on. Some of them did settle in Taylorville and lived here. People sort of shied away from them for a long time. They weren't going to work in the mine, but were going to break Progressive heads at night. Grover knew that something awful was going to happen that night when he saw them and what was on the table by them. They had all sorts of dynamite and everything out at the mine. That's the night they blew up the *Daily Breeze.* Two or three Progressives were beaten rather badly. They did that to get the soldiers here, because John L. Lewis' group was having an awful time. The Progressives were doing pretty well.

The reason they got the soldiers is because, from this corner to the Langleyville road—from that corner clear to the mine on each side as close as they could get—cars were parked and blocking incoming traffic. There were cars in the ditch because they wouldn't let anybody in. That is why they had to blow the place up downtown to get the troops here, so they could stop all this picketing. They weren't working the mine at all; they were at a standstill.

The UMW blew up the *Daily Breeze* themselves. They blamed it on the Progressives. Everybody in town knew that. That wasn't a big secret. No, they did it themselves in order to get the soldiers here. Ray Tombazzi was a very young man and it ruined his life. He went to prison and he was just as gentle, kind, and sweet a guy as you ever want to meet.

I remember the night they bombed Leal Reese's, a PMA attorney, house. There was a house on the corner of Vine Street and they were on Walnut. Anyway, the Reeses lived in that house on the corner. There was a straight shot from our house to theirs across a big wide yard, so when the bomb hit, we got the full blast. Now, Gilbert Large lived next to us, and the Jones boys lived across the street. The bomb blast shook my bed. I was sound asleep and didn't know what had happened. I jumped up to get in the hall, and I could hear my mother say, "I can't find the door." She was so scared she couldn't find the door in her bedroom. We met in the hall and my dad came running out of their bedroom. He was pulling his pants over his pajamas and he was saying, "My God, what did they do? Bomb Reese's house?" I said, "Yes, I guess. I don't know." He said, "I'd better go see if they need any help." So he went out the front door and I rushed out on the front porch because I was worried about him. I didn't know what was going on up there. As he got in front, Gilbert Large was coming out of his house, and the Jones boy and Mr. Jones across the street were coming out of their house. All the men went there together to see if they could help the Reeses.

Leal was a Progressive lawyer, and the people that got blamed were, of course, John L. Lewis' boys, the strike-breakers. They thought if they scared Mr. Reese, he would pull back and wouldn't do anything. The front porch was a shambles. They had one blast then another one came right afterwards. If he had stepped out on the front porch before that second blast went, it would have gotten him; but he didn't. After that I would lie in bed and wonder if they were going to bomb Gilbert Large's house, because by that time they were getting the soldiers in town in a hurry.

When the mine wars came some of those people—swampies—stayed. I hate to say it about some of them, because some are my friends. But some of them had a different type of life, a different type of background, and life was different to them. I am talking about the people they brought up from the southern part of the state. They called them swamp angels. A lot of them stayed and took the place of the miners that could not get their jobs back, or didn't want their jobs back. They were afraid if they went to work they would be killed

Divided Kingdom

Figure 67. Outside the *Daily Breeze* offices, September 28, 1932.

Figure 68. Interior of the *Breeze* building, September 28, 1932. (Courtesy of *Breeze-Courier*)

Divided Kingdom

down below. They would have been, because the feeling was so terrible. They were getting rid of men who were experts and bringing men in who weren't, which was ridiculous. My dad said they were a bunch of low-brow goons.

Sam Taylor

Many Taylorville residents excused the violence as a natural outgrowth of historical antagonisms between miners who were seen as naturally rough-hewn people. As 1935 court records reveal, however, indictments for really violent actions often named certain agent "provocateurs" brought in from the outside by Peabody Coal. Where smaller coal companies signed contracts with the Progressives, there was no violence.

John Cole was state's attorney during most the mine war, and he had one murder trial after another. Generally speaking, these killings were miniature battles, and the local juries apparently took the attitude that, "What the hell, it is a mine war. Why should we try to sort things out for these people? Let them go ahead and kill each other."

The closest I ever came to an actual bombing was in a bootlegging joint in Hewittville out here at the old 58 mine entrance. My cousin and I were in the basement with the two brothers that ran the place, and a bomb went off just outside. It was a hang-out of Progressive miners, who were among other patrons of bootlegging joints in those days.

But, as you probably noticed, there were a number of small bombs. Those small bombs were made by taking about a two and a half gallon can of carbide and putting a couple of inches of carbide in the bottom. Then they poured a bucket of water into it and packed the lid down tight, then run (laughter). They were not lethal, but they could blow out a window very neatly. One of them was put

Divided Kingdom

on the west side of the Baptist church because the Baptist preacher had remarked that it was too bad that the Peabody Coal Company couldn't get along with the local coal miners. That called for "Fats" Cheney's and "Cully" Abrell's duties to go on.

John L. Lewis was considered by most of the coal miners to be hand-in-glove with the Peabody Coal Company. When he signed this contract with the Peabody Coal Company, the Progressive miners organized in Macoupin County, around Carlinville and Virden and towns over there. It never did take fire in the southern Illinois fields. Practically all of the Taylorville miners joined the Progressive union. This was the talk of the community: that Peabody closed some of their mines in southern Illinois and imported the miners into Christian County, and that was a fact. We had over a thousand coal miners move into Taylorville from southern Illinois. At that time, there were lots of them everywhere, and the rooming houses were taking in these folks.

There were lots and lots of new faces, and most of them were in Taylorville because the Midland towns—Kincaid, Bulpitt, Jeiseyville, and Tovey—were simply not safe for the imported miners. Between the National Guard, getting a new sheriff in office in the county, and a new mayor in Taylorville the Progressive Union just petered out.

Incidently, there were either three or four operating coal mines in Pana, and one in Assumption, and one in Moweaqua. They all went Progressive, and the owners of those coal mines just signed up with the Progressives. They never had any battles, no bombings, no killings, no nothing. So most of the people who were not in Peabody management were pretty much sympathetic with the Progressive miners.

I knew the railroad was bombed, and I know that the Progressives were a logical target to accuse. They had the biggest interest in shutting that railroad off, because that would have been a very big blow to Peabody Coal Company.

Divided Kingdom

In fact, apparently it was bombed by amateurs, because they didn't use enough dynamite, or didn't put it the right place. Of course, after that they had a guard stationed there twenty-four-hours-a-day to watch the trestle.

Jesse Lake

Jesse Lake was involved in an ambush at Mine No. 7 in Kincaid on January 3, 1933. Though allied with the UMW, Lake had a very low opinion of the hired "enforcers" for Peabody Coal.

Three or four of us used to ride together to the mine. This one particular night—I think it was about January 3, 1933—we noticed that there was nobody leaving the mine yard, and word got around that the Progressives were going to try to keep us in the mine yard all night. So in the meantime, someone came up with the bright idea, and they cut pieces of cable—I think it was, which served power for either the cutting machine or loading machine—we used it like a billy club. They said, "Well, we're going out and beat the hell out of those guys, so we can get out and go home." So they proceeded to tie white armbands around their arms. That was so you could distinguish me from you out in the dark.

This was in the wintertime, see. I believe we got off at either 4:00 or 4:30 then, and hell, it was nearly dark by the time they got together all that stuff. So we proceeded to go out, and we no sooner got out of that mine gate and all of a sudden it sounded like war broke loose. Boy, you could hear rat-tat-tat-tat and boom-boom, you know, and me and my brother and another guy ran into an old garage, and when it kind of subsided—the shooting—I ran back into the mine yard. When we got out there, those people had a white

Divided Kingdom

Figure 69. Miners' Rally, Kincaid, August 1932. In July and August of 1932 the striking miners held peaceful rallies attended by thousands. Violence was not yet evident in the Midland Tract.

Figure 70. Miners sleeping at rally, Taylorville, August 1932. Many of the rallies lasted days, with miners sleeping on the ground, in cars, and whatever was convenient.

Figure 71. After the bombing of the *Daily Breeze* office in late September, 1932, events turned ugly. More than 10,000 striking miners attended Andrew Gyenes' memorial service in Manner's Park, Taylorville, October 16, 1932. (All courtesy of *Decatur Herald and Review*)

Divided Kingdom

armband around their damned arms, too (laughter)! So you didn't know who was who.

I was just a kid. I weighed about 115 to 120 pounds and when they started shooting, I had enough sense to get back in that mine yard. But I remember one thing distinctly: they started bringing guys back into the mine yard that were shot. I remember one guy. I looked him up in the office after the battle was over and I guess he was Progressive. I'll never forget, he had his shorts down there and he was shot right through both cheeks of his rear. I can remember that just as plain as day. We were scared, because we didn't have any weapons or anything. See, we went out there with just this damn piece of stuff in our hands. And I'm sure there were some Peabody thugs. I'll call them company thugs, and those are the ones that probably had the guns. But after that, we started carrying guns to work, the four of us in the car. I think we had a shotgun and two or three pistols that we carried to work. I thank the good Lord that I never had to use the damn things.

The morning after the battle out at No. 7 mine we were coming to work. Naturally we were scared, and as we went into the mine gate—we used to have to go over a set of tracks, and there was the mine gate right there—there sat a car that belonged to our top boss, name of Hiney Helmer. We found out later that one of the bosses from up in the tipple—a guy by the name of Hickman—was killed in Kincaid. They had been patrolling around the streets in Kincaid, and this one fellow, Hickman, was killed an another guy who was doing some patrolling was name of Fulton Smith; he did not get killed. He was the tipple boss, but the guy that was killed was what they called a table boss. I've always heard that he was a thug. Then I know we had another table boss that was an ex-convict. I know that for sure. He was killed later on in a car wreck. So, evidently, the company brought those guys in or it might be just luck that these people happened to have records or something. But I always thought that they brought those guys in to intimidate the Progressives.

Of course, the Progressives were still active, and I received a copy of the Progressive Mine Workers of America's by-laws through the mail. I thought, "Boy, that's awful damn funny." The next day, riding back from work I said, "You won't guess what? I got a copy of the Progressive Mine Workers by-laws. I think I'll throw the damn thing away. What should I do with it?" This one fellow says, "Read it." I found out later that he was working for both the Progressives and the United Mine Workers. As a matter of fact, he was a mole working for the Peabody Coal Company, but trying to promote the Progressive Mine Workers of America. So that's a little bit of intrigue, I thought. I have never told anybody the guy's name, and I won't now because I'm sure he had people around here. Later on, the man became a boss for Peabody Coal.

Anonymous*

Outrages were committed on both sides. At first, public opinion was favorably disposed towards the Progressives. After four years, however, media, and even public opinion in central Illinois, was more favorable toward the UMW. A progressive explains why the PMA was accused and held accountable for many acts of violence and sabotage.

The hard-core Progressives came from the Gillespie area; that's where it began. Claude Pearcy was the president at that time—first president. The PMA didn't originate in Taylorville; it originated away from here. I think in the Gillespie area, or in Belleville; back in those cornfields. Yes, and then they had the Women's Auxiliary. It was a fight I'll tell you. They had bake sales trying to raise

Divided Kingdom

money to keep the strike going, you know. It took money and they done their share.

When Mrs. Cummerlato was shot [January 3, 1933], the National Guard at the time had Route 48 blocked; you couldn't get here. We got to the funeral, but we took them little country roads. We bypassed the highway, because they had them blocked; you were questioned as to where you were going. They did have martial law, you know. They didn't want us to congregate and they was afraid that funeral might make a large congregation, thereby causing trouble. The Progressives had their Women's Auxiliary, and they were determined they were going to that funeral. I was a driver, and I had four women in my car, and there was another carload behind. There was a lot of news coverage for that funeral.

Figure 72. Casket of Andrew Gyenes in Manner's Park, Taylorville, October 16, 1932. There was a public funeral for every Progressive victim. (Courtesy of *Decatur Herald and Review*)

There was a struggle for public opinion. Well, Peabody Coal tried to get public opinion and they were doing a good job. Public opinion was kind of against the Progressives—your news media. You knew that the bridge was blowed up on C&IM, and suddenly they said, "The striking miners blowed up the bridge." And it wasn't no such thing; they blowed up their own bridge. I talked to Merle Kottle; he was top boss at No. 8 at that time. This was way after World War II when he went to war in the Navy when we talked. We became such good friends, and he says, "My God, I helped load the dynamite in No. 8 when they hauled it up there. You God dang Progressives got blamed for it. I helped blow the damn dynamite."

The man's dead now, but other sources said the same thing—that the dynamite came out of No. 8. You understand these bridges were guarded. There were armed men at each end of the bridge. You think these men are going to blow up that bridge when they got armed guards? That's stupid.

* name withheld at subject's request

Divided Kingdom

Afterword

Maurice Flesher

In 1936 the FBI and the federal judiciary rounded up the PMA miners in Taylorville and other communities they thought were responsible for sabotage of coal company property in Illinois.

One summer afternoon, the word got around that the FBI had arrived in Taylorville and had picked up five Progressive officials concerning the bombing of the Chicago and Illinois Midland Railroad, north of Ball Township High School. I can't remember all the names. One of them was Jack Stanley. There was Henry Noren, as I remember, and his brother. They were taken to Springfield—that would be in the federal court—in the post office building down there.

When they came back to Taylorville, they said that Judge Fitzhugh showed them the evidence that had been compiled against them. It was all piled up on the table, and he wondered what they wanted to do about it. I understood they confessed (laughter). Of course, you will have to look it up to make sure.

See, they had a lot of guards out around all these mines. Even the miners had to guard the railroad trestles on the C&IM. The C&IM paid a second cousin of mine, and I had a friend whose father was a guard in one of the mines. The C&IM paid, as far as I understand. It was my assumption that the C&IM is a [Samuel] Insull Company. Insull owned all of 7, 8, 9, but he did not own 58. 58 was in Taylorville. That is Peabody. And Peabody was merely the coal miner for Commonwealth Edison [Electric Company] and the others.

Tom Rosko

The PMA cause was doomed to failure, since the larger coal companies affiliated with the UMW could monopolize the market and cut costs, while the smaller coal companies allied with the PMA could not compete over the course of time.

Peabody controlled the entire production of coal in Illinois, and that's the reason they wouldn't give in and PMA lost. It was terrible. See, the Progressives could only get more or less the smaller, unproductive, independent mines. They never did organize [the big ones]. Peabody could do as they pleased under the United Mine Workers. They didn't care about conditions or nothing.

The Progressives just picked up Coalton Mine [in Montgomery County] over here and a couple of mines down in the South. They were more or less independent mines. Very seldom did they pick up companies that had three or four or five mines. So they picked up some of the miners that were local. You take Rice Miller, the owner of Coalton Mine. He didn't want to go to the United Mine Workers to negotiate because he figured, well, I don't want no trouble with my workers because of the United Mine Workers. At one time we had about 18,000 miners, or maybe about 20,000, in the area but now they've got nothing. They're lucky if they've got a thousand.

As time went on, the PMA lost strength. Every time a mine shut down that's what you lost. I think they was only in existence about sixteen years. An then the union just got, you know, wasteful, just like the rest of them.

Now, the Progressive mines are all shut down. We were all United Mine Workers at one time, and we pulled away. Up here [in Taylorville] and Witt it was always Irish, English, and Slavic, and stuff like that. We all pulled away to the Progressives. There must have been only about six or eight families from Witt that stayed with the UMW.

Ray Tombazzi

Ray Tombazzi's post-mortem provides the best picture of how the PMA gradually faded away.

The war faded away gradually. A lot of miners went back to work, and I talked to them and told them it was no use; we weren't making no headway in the union. It seemed like it was a status quo deal, and it seemed like this was as far as we could go. They tried to go into Pennsylvania. They tried in Kentucky; we tried in Virginia and West Virginia. I would organize them little mines, and they would impose such restrictions on them that they couldn't compete with big coal companies. They would go out of existence, so I quit organizing.

We won the war. Principally and morally we had the populace with us, but financially we didn't make it. The Springfield press was favorable. The consensus of opinion among all the populace was favorable to our cause; no question about it. But the UMW with their finances overwhelmed us.

I'd say there was collusion between the operators and John L. Lewis at that time. I'd say the operators kicked in, too, to help Peabody because they were fighting the same battle. It was an economic battle, so to speak, because mechanization was coming in on them and technology was moving fast. To put it bluntly, social economy wasn't keeping up with the industrial economy.

Conclusion

The "Miners' War", "Community Based" Unionism and the Formation of the CIO

The memoirs recorded here cast light on the many hidden recesses of life in one of the country's premier coal fields during a period of turmoil. The interviews indicate that the miner's insurgency was triggered specifically by Peabody Coal's and the UMW's move toward the consolidation of their power. More specifically, the PMA's prolonged strike was a response to several factors in Peabody's reorganization of their mines including the use of arbitrary power, mine mechanization, the deskilling of the miner's job, and the resulting massive lay-offs.

The strike's ultimate thrust and direction, however, was firmly shaped by the workers whose values and objectives emerged from a distinct tradition of community solidarity which gave the new union a radically different focus. The testimonies recorded in **Divided Kingdom**, for example, describe a kind of union qualitatively different from the bureaucratic business unions that came to dominate the CIO in the mid and late 1930s. In fact, other recent studies

Divided Kingdom

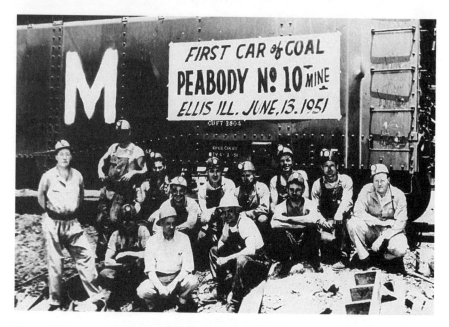

Figure 73. After the strike, Peabody Coal Company became the most prolific producer of coal in the world. Mine No. 10 between Pawnee and Tovey was the world's largest coal mine when it opened June 13, 1951. (Courtesy of the Boch Brothers)

of labor in the early 1930s, which also relied on the testimonies of rank-and-file, offer evidence that these early efforts at organizing egalitarian unions preceeded the formation of the CIO.

This early unionism has been called "community based" or "solidarity unionism" and emphasized local autonomy and opposed the kind of bureaucratic unionism represented by the UMW and the model John L. Lewis would introduce in the new CIO. This alternate unionism was democratic, deeply rooted in mutual aid among workers at different work sites, and independent of New Deal politics. For example, in contrast to the UMW, miners in strong grass roots unions in Illinois favored schemes to share and "equalize" the work among all workers **regardless of seniority**.

Divided Kingdom

Historians have assumed that the general strikes in the Northern states and the miners' strikes in Pennsylvania and Illinois in the early 1930s were isolated events. Quite the contrary, the local strikes were symptomatic of the mobilization of working-class communities. In the absence of effective and concerned national organizations from which they could seek help, rank and file miners and other workers were obliged to turn to each other and create horizontal networks that in turn generated a distinctive organizational culture and set of attitudes.

Locally, the self-organization of rank-and-file miners was at least as effective as the top-down efforts of the UMW. In the fall of 1932, for example, the Illinois Department of Mines reported that there were 60,000 PMA members in Illinois and only 10,000 UMW miners. This new unionism was effective in other ways as well. PMA sit-down strikes in Wilsonville, St. Claire, and Gillespie in the mid 1930s were spontaneous and resulted in immediate victories. In eastern Pennsylvania, a general strike in the Anthracite region and the "bootlegging" of coal to Philadelphia in order to raise money for the locals of the Anthracite Workers of America led to recognition and a modified acceptance of job sharing. And the celebrated sit-down strike in Flint, Michigan in 1936, copied from the example of the PMA and the Anthracite Miners, was effective in ringing broad concessions from General Motors and Fisher Body.

It is equally as clear that the community-based, horizontal "culture of struggle" in Illinois coal fields had its roots in a tradition of community-based strikes against the coal companies in the teens and 1920s and a desire to reform the UMW and its bureaucracy. The PMA represented only the latest effort in a long line of failed local attempts to organize a new union. For example, the Reorganized United Mine Workers and the National Miners Union, with some successes in Illinois in the 1920s and early 1930s, attempted to implement some of the same organizational responses as the PMA later did to threats to steady employment and traditional working arrangements. Reinstituted rank and file representation in new unions were a constant response to overt oppression and UMW indifference in the 1920s and an indigenous move towards economic demands and representation.

Initially, the Progressive Miners of America were not just an accommodation to monopoly capitalism, but represented a whole realm of ideas which would make coal mining more humane. Skilled miners, after all, had made most of the innovations in coal mining technology in the central Illinois coal fields and had spoken at length about the need to implement new technologies, and about the need for decent working conditions, joint labor and capital planning, and the need to implement new technologies which took the miners' needs into account.

In 1932 these hopes did not seem naive. It did not appear inevitable that the only way the new technologies could be implemented would be destructive to the miners, nor was it clear then that the PMA could not compete with the power of the UMW. Major PMA leaders such as Joe Ozanic, Roy Tombazzi, and Jack Batuello felt it was possible to install major new technologies under PMA sponsorship with little negative impact on the miners.

In hindsight, however, it was inevitable that certain technologies would emerge and that the business oriented unionism of John L. Lewis and the UMW would prevail over the wishes of the rank and file. We now know John L. Lewis' collusion with Peabody Coal made it impossible even to consider a particular PMA strategy; we also know that Peabody Coal needed the Lewis partnership to reduce the work force because of the competitive pressure it was under; and that John L. Lewis, in turn, needed Peabody Coal to consolidate the UMW's power in the troublesome Illinois coal fields. In short, the bargain struck between Peabody Coal and Lewis doomed any serious discussion of a PMA-negotiated plan to develop new technologies, and therefore soften the blow of massive cutbacks in manpower.

As the 1930s wore on, the miners themselves became aware of their impending defeat and the need to accommodate the new order as advocated by Peabody Coal and the UMW. Interwoven throughout the interviews were expressions of the need for community and family security; the desire to fight for job security; decent and effective union representation: and limits on the abuses of power caused by mechanization. It is an American tragedy that reasonable people did not hear them and that American labor was forever saddled with the bureaucratic unionism of John L. Lewis.

Appendices

Appendix A
Glossary

Air shaft: The air hole that reaches from the surface of the mine to the bottom. Without this, breathing air could not be circulated throughout the mine.

Black damp: Pockets in the underground mine in which there is no oxygen.

Black lung, or miners' asthma: A disease suffered only by underground coal miners. Their constant inhalation of coal dust chronically inflames the lungs, causing shortness of breath, premature fatigue, and, often, death. The incidence of black lung increased in the twentieth century because of the great amount of coal dust generated by the new mining machinery.

Breast auger: A hand drill strapped around the miner's chest used to bore a hole in the coal for explosives.

Buggy: A coal car.

Cage: An elevator that hoisted coal cars in and out of the coal mines. Miners rode in them to descend and ascend out of the coal mine.

Divided Kingdom

Carbide lamp: A miner's lamp that fit on to the canvas cap of early twentieth century coal miners. The bottom part of the lamp held carbide and upper contained water, which would drip on the carbide and release gases that would be ignited by combustion to cause light.

Clean the coal: Hand loaders underground had to pick rock out of the coal before they shovelled it into coal cars. If machine-loaded, rock pickers on top had to pick the rock out.

Cooperative store: A general merchandise and grocery store run by subscription to common stock. In the coal fields, specific ethnic groups ran their cooperatives.

Cross cut: After the miners cut rooms into the coal for fifty feet, horizontal cuts were made across the rooms to provide openings for air.

Deskilling: A move in industry to take a complex, skilled job and reduce it to minute and specialized tasks that can be performed by unskilled laborers.

Dirt gang: A crew of mine workers that cleans up rock falls at the entries of rooms, retrieves coal from spills, and cleans track.

Dock boss: A mine manager, usually positioned on the tipple, who fines or "docks" miners for having too much rock in loads of coal.

Face: The wall that fronts on the haulage line where the miner begins to dig into a thirty foot wide room.

Five BU: The first mechanized loading machine manufactured by Joy Manufacturing. This small loading machine collected coal, loaded it on a conveyor belt, and then emptied into a coal car.

Hand loader: A mine worker who shovelled coal into a coal car for the miner.

Haulage lines: Corridors in underground mines through which coal was moved by mules, electric motors, or conveyor belts.

Lifter shots: A shot placed in the face about two feet off the floor that lifted the coal upward on explosion.

Little dog mines: A small mine developed with little capital and few employees. Miners considered these mines dangerous.

Motors
    Inside motor: electric car that gathered coal cars together for trip to the cage.

    Main line motor: electric car that brought empty coal cars to the miners.

    Man trip: electric car with empties returning to the face. Often, miners would ride in the empties to their places underground.

Motor men: Drivers who steer the electric motor.

Number 2 banjo: Number 1 and Number 2 banjo refer to the size of the shovel miners used. The Number 2 was larger and took more strength to handle.

Divided Kingdom

On the level: When coal seams ran horizontally underground the miner would "work on the level," i.e., standing straight up. In some coal fields, the coal seam was pitched, i.e., not "on the level," and harder to mine.

On the solid: The drilling or blasting of coal with no undermining of the seam.

Picking tables: Extended benches where young boys picked rock from coal.

Pit committee: The union committee of one to three men that solves miners' grievances. The committee is the key representative of the miner.

Recovery gang: Gang of underground laborers who recovered track near the face so it could be used in other areas of the mine.

Rib of coal: Two lines of coal or ribs of coal appeared on the left and right sides of a "room" as the miner dug further into the face.

Rock pickers: Coal companies employed young boys 13-17 to pick rock out of the coal.

Roof bolts: Fasteners that held up the ceiling of the underground mines to prevent rock falls. They took the place of props.

Room neck: The first space out of the entry. Off the room neck every fifteen feet there was another room dug into the face.

Divided Kingdom

Room and pillar mining: A method of mining in which each miner had his individual 30' x 20' room to mine out. Longwall mining eventually replaced the method.

Shot firer: Company man who ignited the shots bored in the holes in the face. In Illinois mines shot firers fired on the night shift when miners were gone.

Snubber: A short fuse shot; a dowel rod with little powder.

Stickers: A marker, an I.O.U., or credit slip usually drawn on the coal company's store so the miner could buy supplies.

Territory: Sixteen rooms with a cross cut constituted a territory.

Timberer: Company man who placed props in the underground mines to hold up the ceiling or top.

Tipple: The roof or ceiling of the underground coal mine. Roof cave-ins were the major causes of death underground, so the miners wanted to "sound" their roof or top before mining coal.

Trapper or Door man: Company man who opened the wooden doors at the entrance of the mines so motors could exit without stopping and closed them so fresh air could circulate throughout the underground workings.

Trip rider: Men who rode on the trips up front and threw the switch for the motor men, coupled coal cars, and opened the wooden "trap" doors. These men were injured by getting squeezed between coal cars.

Divided Kingdom

Appendix B
Questionnaire
Christian County Coal Miner's Project

1. Basic information

I am _____. This interviewing project is jointly sponsored by the Illinois State Historical Society and the Illinois Historic Preservation Agency. Here enter the date, subject's name, present address, year of birth, and birth place.

2. Household - parents:

How many years did you live in the house where you were born? Where did you live then? (Continue for moves up until you left home.) Do you remember why the family made these moves? (If immigrants, probe more about the journey, economic effect of the move, continuing contact with original home, and other migrants.)

3. Household - routine:

I should like to ask you now about life at home when you were a child. How were the rooms used? Bedroom, other rooms; furniture. Did anyone else besides your parents and brothers and sisters live in the house? Other relatives, or lodgers? (If lodgers, where did they sleep? Where did they eat? How much did they pay?) How did the housework go? Who made or mended the family's clothes? Were any clothes bought new or secondhand? Where and when? What did your dad do around the house? Did he help your mother with any of the jobs in the house? Did you have any tasks you had to carry out regularly at home to help your mother or father? How long did you continue to do these tasks? Probe for brothers and sisters. Did you have a particular bedtime? Who put you to bed? Sleeping arrangements of the whole family. How did family manage with washing and bathing.

4. Meals:

What did you usually eat and drink? Did you have anything different on certain days (Sunday)? Did your mother and father bake bread, make jam, butter, or can fruit or vegetables, make pickles, wine, beer, or any medicine for the family? Did your father or mother grow vegetables and fruit? Did they keep any livestock for the family (pigs, chickens, goats)?

Did you supplement your diet with game?

What was your parents' attitude if you left some food uneaten on your plate? Did all the family sit at the table for the meal?

5. General relationships with parents:

Were your mother or father easy to talk to? Did she show affection? If you had any worries could you share them with her or not? (Repeat for father.)

Divided Kingdom

How were you expected to act around parents? As a child, was there any older person you felt more comfortable with than your parents? (Grandparents, other relatives, servants.)

What kind of person do you think your parents hoped you would grow up to be? Did your parents bring you up to consider certain things important to life? If you did something that your parents disapproved of, what would happen? (for example, swearing.) If punished: by whom; how often? Do you remember any particular occasion on which you were punished?

6. Religion:

How did you spend Sundays? Did your parents think it was wrong to work or play on Sunday? Did your parents attend church? Denomination? How often? Both mother and father? Did they hold positions in the church? Did you attend?

Did you go to Sunday School or not? Choir? Temperance meetings? Evening classes? Other activities?

Was grace said at meal in your family? Prayers? How much would you say religion meant to you as a child?

7. Marriage:

What age were you when you married? How long had you known your wife then? How did you meet? Where did she come from? What kind of family? Did you save up money before getting married or not? Did your parents help you in setting up a home? Did they help you later on? Or did you help them? Could you describe the wedding? Did you have a honeymoon? Where did you live after you married? (Did you ever consider moving out of the area when you first married?) Where did you live then?

8. Children:

Did you have any children? How many? Names and years of birth. Relationship with each. Were your children born at home?

9. Family life after marriage:

Budget and control of household: I want to ask you how you and your wife managed the household in those days, especially the Depression period. How much of your earnings would you give to your wife at the time? Did you pay your bills promptly, or slowly—and did you do it yourself? Did you buy new furniture; clothes for the children; go on outings and vacations? Who looked after the garden? Rephrase and use Sections 3, 4, and 6 again now.

10. Community, social class, and neighborhood:

Did anyone outside the home help your wife look after the house or family? (Relations, friends, neighbors.) In what ways? How often? If your wife was ill or confined to bed how did she manage? Do you remember what happened when one of your children was born?

What relations lived nearby? When did you see them? Where? Do you remember them helping you in any way, teaching you anything? Did you have friends? Where did they live? Where did you see them? Did your wife have friends of her own? Where did she see them? Did she visit anyone who was not a relation?

Divided Kingdom

Were people ever invited into your home? How often? Who were they? Would they be offered anything to eat or drink? On any particular days or occasions? Would you say that the people invited in were your friends or your wife's friends, or both? Did people call in casually without an invitation? When? People often tell us that in those days they made their own amusements. What did you do when you got together with your friends/neighbors? (Music, games.) Many people divide society into different social classes, or groups. In that time during the Depression did you think of some people belong to one and some to another? Could you tell me what the different ones were? What class/group (Informant's own term) would you say you belonged to yourself? What sort of people belonged to the same class/group as yourself? What sort of people belonged to the other classes/groups you have mentioned?

Can you remember being brought up to treat people of one sort differently than people of another sort? Did you ever touch your hat; show respect in some way? To whom? Was there anyone you called "sir" or "madam"? Do you remember anyone showing respect to you in those days? In the town, who were considered the most important people? Did you come into contact with them? Why were they considered the most important? If respondent middle or upper class: Would these people have been considered at the time to be "in society"?

Where you lived, did all the people in the working (or lower, or other term used by informants) class have the same standard of living, or would you say there were different groups? Describe a family within each group. Do you think that one group felt itself superior to the rest? Were some families thought of as rough, and others as respectable? Do you remember a distinction of this kind between craftsmen and laborers?

Divided Kingdom

Was your home rented? If yes: What do you remember of the landlord? Did you belong to any savings club? (Insurance, sick, funeral, etc.) Do you remember feeling that you had to struggle to make ends meet? (If not: did you help poorer people in any way? Did you belong to any philanthropic organizations?) What difference did it make to the family when you were ill or out of work? Did you ever get help from the church or any charity? How did they treat you? How did you feel about that? Any help from New Deal Programs: CCC, the WPA, etc.

11. The workplace:

Describe the actual workplace where you performed work—did it change; actual work you did (describe the process); for how long; then what; did you like your work; did your work process change at all; did you notice that your supervisor came to supervise you more closely; did the role of foreman change, especially after the 1920s (have an in-depth explanation if possible; who did the hiring, wages, set production quotas); did he show favoritism, abuse his position, changes in method pay (wage pay as opposed to incentive pay—percentage piece work; if possible describe the different jobs available—piece work and straight time); skilled work being phased out; was there an increase in production quotas especially in the mid 1920s and 1930s; were contracts with workers simple or verbal; was management aware of workers' rights?

Composition of the workforce: percentage of Lithuanians, Italians, Polish, Slovak, etc.; turnover of the population, especially in the 1920s and 1930s; did they all get along; how did you get on with the other people you worked with; could you talk or relax (did you play cards, etc. in the breaks?) What was your relationship to neighborhood

Divided Kingdom

groups; did workers feel a solidarity among themselves? Did any of the workers visit fellow workers at times of illness or death, normal times? Were there distinct neighborhoods where workers or groups of workers from the mine lived? Same groups as in Section 10?

How did your employer treat you? How did you feel about him? If there was a worker dispute before the union, were there such things as stoppages, slowdowns, informal negotiations? Did abuses and favoritism cast doubt about the effectiveness of the UMW unions? Why the PMA union? Who organized the PMA union? Was there an organizing committee? Who was on it? Was there a PMA cadre in the mine-ethnic composition, work departments, skilled outside help.

Strike of 1932 and Mine War: Describe the strike; who organized it, town's reaction; state militia's and mayor's role; did it spread; issues (what was disputed?) How resolved?

Pit Committee: Foreman and their roles; relationship to workers especially in the 1920s and 1930s; issues-grievance procedures; seniority; general strikes; wildcat strikes; opinion of union now.

12. The future:

What have you learned from your life, neighborhood, and work experience? Was it a desirable life? What kind of community do you want? Housing, neighborhoods, work, etc.?

Short Bibliography
On Labor and Oral History

Booth, Stephanie E. "Christian County Mine Wars, 1932-1935" (unpublished paper available at Taylorville, Illinois, Public Library), 1977, 16pp.

_____. "Gerry Allard: Miner's Advocate" (unpublished Master's Thesis, Illinois State University), 1972.

Bodnar, John. *The Transplanted: A History of Immigrants in Urban America*. Bloomington, Indiana: Indiana University Press, 1986.

Brecher, Jeremy, Jerry Lombardi and Jan Stackhouse. *Brass Valley: The Story of Working Peoples' Lives and Struggles in an American Industrial Region*, Philadelphia: Temple University Press, 1982.

_____. *Preserving Workers' History: A Guide* (Derby, Connecticut: Brass Workers History Project), n.d.

Broverman, Helen B. and Dorothy D. Drennon, ed. *History of Christian County, State of Illinois*. Illinois Sesquicentennial Edition. Jacksonville, Illinois: Prod. Press, Inc., 1968.

Davis, Cullom. *From Tape to Type*, Chicago: American Library Association, 1975.

Dix, Keith. *Work Relations in the Coal Industry: the Handloading Era*, 1880-1930, Morgantown, West Virginia: Institute for Labor Studies, 1954.

Grele, Ronald J. "Movement Without Aim: Methodology and Theoretical Problems in Oral History," in Grele, ed. *Envelopes of Sound*, Chicago, Illinois: Precedent Publishing, Inc., 1975.

Harevan, Tamara. "The Search for Generational Memory: Tribal Rites in Industrial Society," *Daedalus*, (Fall) 1978.

Hudson, Harriet Dufresne. *The Progressive Mines of America*, 1932-1946. Chicago: The University of Chicago Press, 1950.

Krohe, James Jr. *Midnight at Noon: A History of Coal Mining in Sangamon County*. Springfield, Illinois: Sangamon County Historical Society, 1975.

Lantz, Herman R. *People of Coal Town*, New York: Columbia University Press, 1958.

Laslett, J.H.M. "Swan Song or New Social Movement? Socialism in Illinois District 12, United Mine Workers of America, 1919-1926," in Donald T. Critchlow, ed., *Socialism in the Heartland: the Midwestern Experience*, 1900-1925, Notre Dame: Notre Dame Press, 1986.

Oblinger, Carl. *Interviewing the People of Pennsylvania: A Conceptual Guide to Oral History*, Harrisburg: Pennsylvania Historical and Museum Commission, 1981.

_____. "The Coal Union War: How the Progressives Fought John L. Lewis," *Illinois Times*, Vol. 10, No. 32 (April 11-17, 1985), 4-10.

Young, Dallas, *A History of the Progressive Miners of America*, 1932-1940. Urbana, Illinois: University of Illinois Press, 1940.